The phenomenon of electric and magnetic field vector (wave) propagation through the free-space, or any other medium is considered within the ambit of electromagnetics. The media themselves, in general, could be of diverse type, such as linear/non-linear, isotropic/non-isotropic, homogeneous/inhomogeneous, reciprocal/non-reciprocal, etc. Such electromagnetic wave propagation problems are formulated with the set of Maxwell's equations. Computational Electromagnetics endeavors to provide the solution to the Maxwell's equations for a given formulation. It is often difficult to find closed forms solutions to the Maxwell's equation formulations. The advent of computers, and particularly the initial developments of efficient coding for numerical analysis, encouraged the development of numerical electromagnetics. A second motivation came from the interaction of the electromagnetic wave with the matter. This could be visualized as scattering bodies, which required incorporation of the phenomena of reflection, refraction, diffraction and polarization. The finite/large nature of the scatterer required that problem of electromagnetics is considered with respect to the operation wavelength leading to the classification of low-frequency, high-frequency and resonance region problems. This also inspired various asymptotic and grid-based finite-method techniques, for solving specific electromagnetic problems. Surface modeling and ray tracing are also considered for such electromagnetic problems. Further, design optimization towards hardware realization have led to the recourse to various soft computing algorithms. Computational Electromagnetics is deemed to encompass the numerical electromagnetics along with all other above developments. With the wide availability of massively parallel high performance parallel computing platforms, new possibilities have emerged for reducing the computation time and developing macro models that can even be employed for several practical multi-physics scenarios. Both volume and surface discretization methods have been given a new boost, and several acceleration techniques including GPU based computation, learning based approaches, and model order reduction have been attempted. Limitations of generating meshes and modifying these for parametric estimation have been addressed by statistical approaches and smart solvers. Many nature-inspired algorithms and other soft computing approaches have been employed for electromagnetic synthesis problems. One of the recent additions is Game Theoretic optimization.

Finally, the emergence of Computational Electromagnetics has been motivated by myriad applications. These diverse application include but are not restricted to those in Electronics and Communication, Wireless Propagation, Computer Hardware, Aerospace Engineering, Biomedical Engineering, Radio-astronomy, Terahertz Technology, Photonics, etc for modelling of devices, components, systems and even large structures. Some of the well-known applications are in Analysis and design of radio frequency (RF) circuit, antennas and systems, Analysis of antenna on structures, radar imaging, radar cross section (RCS) computation and reduction, and analysis of electromagnetic wave-matter interactions at discrete, random and periodic geometries including metamaterials. Authors are encouraged to submit original research work in the area of Computational Electromagnetics. The content could be either theoretical development, or specific to particular applications. This Series also encourages state-of-the-art reviews and easy to comprehend tutorials.

More information about this subseries at http://www.springer.com/series/13885

Vineetha Joy · G. L. Rajeshwari ·
Hema Singh · Raveendranath U. Nair

Fundamentals of RCS Prediction Methodology using Parallelized Numerical Electromagnetics Code (NEC) and Finite Element Pre-processor

 Springer

Vineetha Joy
Centre for Electromagnetics (CEM)
CSIR-National Aerospace Laboratories
Bengaluru, Karnataka, India

Hema Singh
Centre for Electromagnetics (CEM)
CSIR-National Aerospace Laboratories
Bengaluru, Karnataka, India

G. L. Rajeshwari
Centre for Electromagnetics (CEM)
CSIR-National Aerospace Laboratories
Bengaluru, Karnataka, India

Raveendranath U. Nair
Centre for Electromagnetics (CEM)
CSIR-National Aerospace Laboratories
Bengaluru, Karnataka, India

ISSN 2191-8112   ISSN 2191-8120   (electronic)
SpringerBriefs in Electrical and Computer Engineering
ISSN 2365-6239   ISSN 2365-6247   (electronic)
SpringerBriefs in Computational Electromagnetics
ISBN 978-981-15-7163-3   ISBN 978-981-15-7164-0   (eBook)
https://doi.org/10.1007/978-981-15-7164-0

This Springer imprint is published by the registered company Springer Nature Singapore Pte Ltd.
The registered company address is: 152 Beach Road, #21-01/04 Gateway East, Singapore 189721, Singapore

To

Our Families

Preface

With the advent of stealth technology, precise computation of radar cross section (RCS) has become an inevitable component in the efficient design and development of military vehicles. However, the scattering characterization of stealth platforms is undoubtedly the most demanding problem in modern applied electromagnetics. When it comes to the computation of scattered fields from conducting structures, the accuracy and efficacy of a method of moments (MoM)-based integral equation formulation are difficult to be surpassed. Among MoM-based solvers, Numerical Electromagnetics Code (NEC) is a versatile open-source computer program used for the electromagnetic analysis of metallic structures in the presence of sources or incident fields. Since its development, NEC has continued to be one of the most widely used electromagnetic simulation codes even in the presence of commercial MoM-based solvers. This can be attributed to the availability of the well-documented computational engine. However, a serious drawback of NEC is the absence of an appropriate module for the wire-grid-based meshing of geometries and the consequent generation of segmentation data in an NEC compatible format. This is one of the most intricate and time-consuming steps in the computation of RCS using NEC. In this regard, this brief presents a detailed methodology for the computation of RCS of metallic structures using a parallelized version of NEC in conjunction with a finite element pre-processor, which has been strategically incorporated for geometry modelling catering to NEC guidelines. It includes a thorough overview of the theoretical background of NEC including all relevant aspects of formulation and modelling. This book will serve as a valuable resource for students, researchers, scientists, and engineers working in the field of RCS predictions and measurements.

Bengaluru, India

Vineetha Joy
G. L. Rajeshwari
Hema Singh
Raveendranath U. Nair

Acknowledgements

First and foremost, we would like to thank the God Almighty for being the beacon of hope during the course of this research work.

Further, we would like to thank Shri. Jitendra J. Jadhav, Director, CSIR-National Aerospace Laboratories, Bengaluru, for the permission to write this SpringerBrief.

We would also like to acknowledge the valuable suggestions from our colleagues at the Centre for Electromagnetics during the course of writing this book.

But for the concerted support and encouragement of Ms. Swati Meherishi, Editorial Director, Applied Sciences and Engineering, and Ms. Muskan Jaiswal of Springer Nature India Private Limited, it would not have been possible to bring out this book within such a short span of time.

We are also forever beholden to our family members for their incessant inspiration which supported us to stay at rough tracks. Vineetha Joy would like to specially thank her husband Johnu George for his constant support and encouragment during the course of this work.

Contents

About the Authors

Mrs. Vineetha Joy is working as Scientist at Centre for Electromagnetics of CSIR-National Aerospace Laboratories, Bangalore, India since March, 2016. She obtained B. Tech in Electronics and Communication Engineering with Second Rank from University of Calicut in 2011 and received M. Tech degree with First Rank in RF & Microwave Engineering from Indian Institute of Technology (IIT), Kharagpur in 2014. She is actively involved in various R&D programs spanning domains like Computational Electromagnetics, Electromagnetic Design and Performance Analysis of Radomes, Design and Development of broadband radar absorbing structures (RAS), Hybrid Numerical Techniques for Scattering Analysis, and Electromagnetic Characterization of Potential Materials for Airborne Structures. She has authored/co-authored several research publications including peer reviewed journal papers, symposium papers, technical documents and test reports.

Ms. G. L. Rajeshwari obtained M. Tech. in Digital Electronics and Communication in 2019 from Department of Electronics & Communication Engineering, Bangalore Institute of Technology, Bengaluru, India and B.E. in Electronics & Communication Engineering, in 2017 from Nitte Meenakshi Institute of Technology, Bengaluru. Her research interest includes Computational Electromagnetics, Hybrid Numerical Techniques for Scattering Analysis etc.

Dr. Hema Singh is working as Senior Principal Scientist in Centre for Electromagnetics, National Aerospace Laboratories (CSIR-NAL), Bangalore, India. She has received Ph.D. degree in Electronics Engineering from IIT-BHU, Varanasi India in Feb. 2000. For the period 1999-2001, she was Lecturer in Physics at P.G. College, Kashipur, Uttaranchal, India. She was a Lecturer in EEE of Birla Institute of Technology & Science (BITS), Pilani, Rajasthan, India, for the period 2001-2004. She joined CSIR-NAL as Scientist in January 2005. Her active areas of research are Computational Electromagnetics for Aerospace Applications, EM analysis of propagation in an indoor environment, Phased Arrays, Conformal Antennas, Radar Cross Section (RCS) Studies including Active RCS Reduction.

She has contributed in the projects not only sponsored by DRDO on low RCS phased array, Active RCS reduction, RAS development, but also in the project sponsored by Boeing USA on EM analysis of RF field build-up within Boeing 787 Dreamliner, She received Best Woman Scientist Award in CSIR-NAL, Bangalore for period of 2007-2008 for her contribution in the area of active RCS reduction. Dr. Singh co-authored 14 books, 2 book chapters, 7 software copyrights, 370 scientific research papers and technical reports.

Dr. Raveendranath U. Nair is working as Senior Principal Scientist & Head in Centre for Electromagnetics, National Aerospace Laboratories (CSIR-NAL), Bangalore, India. He holds a Ph.D. degree in Microwave Electronics from MG University, India. He has over 20 years' experience in the field of electromagnetic design and analysis of radomes, design and development of FSS based structures for airborne platforms, radar cross section (RCS) studies, design and development of artificially engineered materials etc. He has contributed significantly to the National Radome Programs including Doppler Weather Radar (DWR) radome installed at SHAAR Sreeharikota, Fire Control Radar (FCR) radome for Jaguar aircraft, Astra missile ceramic radome, nosecone radome for Saras aircraft, multi-band radomes for TU-142M aircraft etc. He has authored/co-authored over 200 research publications including peer reviewed journal papers, symposium papers, technical reports, and two books.

List of Figures

List of Tables

Chapter 1
Introduction

The burgeoning research and developmental activities happening in various domains like antenna design, wireless communication, stealth technologies, electromagnetic interference/compatibility (EMI/EMC), etc., have made systems much more complex than ever before. Considering the huge expenses related to fabrication, it is not practically possible to modify a device if its performance is not within the stipulated limits. Therefore, techniques capable of precise characterization of complex systems are the need of the hour. Closed-form analytical solutions are available for only particular problems, which can hardly be directly applied in real-world systems. This gap has led to extensive research on computational electromagnetics (CEM) where numerical methods are used for solving electromagnetic problems. Among all the numerical techniques, method of moments (MoM) is highly accurate and efficient especially for the analysis of scattering from conducting structures. In this regard, Numerical Electromagnetics Code (NEC) is a versatile MoM-based open-source code (written in Fortran) used for the electromagnetic analysis of metallic structures in the presence of sources or incident fields (Burke and Poggio 1981; Burke et al. 2004). The development of NEC has been spearheaded by Lawrence Livermore Laboratory, USA, with the support of the Naval Ocean Systems Center and the Air Force Weapons Laboratory.

Since its development, NEC has continued to be one of the most widely used electromagnetic simulation codes even in the presence of commercial MoM solvers. This attributes to the availability of the well-documented powerful computational platform, which can be used as a foundation for further developments in the area of electromagnetic analysis. NEC does not use many of the simplifying assumptions employed by other high-frequency asymptotic techniques and thus provides a highly accurate and resourceful tool for electromagnetic studies.

The numerical solution of integral equations, for the induced currents on metallic structures by sources or impinging electromagnetic waves, is the essence of NEC. The excitations can be defined in terms of voltage sources or plane waves of linear/elliptical polarization. Various electromagnetic entities like non-radiating

© The Author(s), under exclusive license to Springer Nature Singapore Pte Ltd. 2021
V. Joy et al., *Fundamentals of RCS Prediction Methodology using Parallelized Numerical Electromagnetics Code (NEC) and Finite Element Pre-processor*,
SpringerBriefs in Computational Electromagnetics,
https://doi.org/10.1007/978-981-15-7164-0_1

networks/transmission lines, perfect/imperfect conductors, perfectly or imperfectly conducting ground planes, etc., can be easily included in NEC. The output of NEC can be tailored for specific requirements like induced current and charge densities, near magnetic or electric fields, impedance or admittance, radiated fields, etc. Furthermore, many of the commonly encountered performance parameters such as gain and directivity are also available. Therefore, the program is well suited for both antenna analysis and radar cross section (RCS) predictions (Burke and Poggio 1981).

Although numerical methods like MoM can be used in a wide range of radiation and scattering problems, the computational complexity increases with increase in the electrical size of the structure under consideration. The memory requirements have been found to exhibit a quadratic growth with increase in the number of segments. A parallelized version of NEC has been introduced (Rubinstein et al. 2003, Rubinstein 2004) to tackle this problem. The two most time-consuming steps in evaluating the response of a structure, viz. (i) computation of the interaction matrix elements and (ii) solution of the resultant matrix equation, have been efficiently parallelized in Rubinstein et al. (2003).

In addition to the memory constraints, a serious issue in the computation of RCS using NEC is the wire-grid-based meshing of geometries and the generation of NEC input file with the segmentation data included in accordance with NEC guidelines. This is an extremely difficult and error-prone process, especially for complex structures (Hubing et al. 1994; Toit and Davidson 1995; Ross et al. 1999), and furthermore, the code will often run in the presence of such errors and produce results that are plausible but incorrect. Therefore, an appropriate pre-processor is essential for the accurate prediction of RCS using NEC.

The aim of this book is to provide a strong foundation for the computation of radar cross section (RCS) of aerospace structures using a parallelized version of NEC in conjunction with a finite element pre-processor. This book has five chapters. It begins with a thorough overview of the theoretical background of NEC including all relevant aspects of formulation of integral equation and its numerical solution. The strategy used for parallelization of the code is also included in Chap. 2. The geometrical modelling of structures and the input/output formats are described in Chap. 3. The revised methodology for the computation of RCS incorporating all the required steps and details is discussed elaborately in Chap. 4 followed by case studies and validations in Chap. 5.

References

Burke, G.J., and A.J. Poggio. 1981 January. Numerical Electromagnetics Code (NEC)—Method of Moments, Lawrence Livermore National Laboratory, Livermore, California, Technical Document, Rep. UCID-18834, 719.

Burke, G.J., E.K. Miller, and A.J. Poggio. 2004 June. "The Numerical Electromagnetics Code (NEC)—A Brief history." In *Proceedings of IEEE Antennas and Propagation Society Symposium*, vol. 3, 2871–2874, Monterey, California.

Hubing, T.H., C.H.H. Lim, and J. Drewniak. 1994 March."A Geometry Description Language for 3D Electromagnetic Analysis Codes." In *Proceedings of the 10th Annual Review of Progress in Applied Computational Electromagnetics*, Monterey, CA, 417–422.

Rubinstein, A., F. Rachidi, M. Rubinstein, and B. Reusser. 2003. A Parallel Implementation of NEC for the Analysis of Large Structures. *IEEE Transactions on Electromagnetic Compatibility* 45 (2): 177–188.

Ross, J.E., L.L. Nagy, and J. Szostka. 1999. "CAD tools for vehicular antennas." In *Proceedings of the 1999 IX National Symposium of Radio Science*, Poznan, Poland, March 16–17.

Rubinstein, A. 2004 January. Simulation of Electrically Large Structures in EMC Studies: Application to Automotive EMC, Ph.D. dissertation, École polytechnique fédérale de Lausanne (EPFL), https://doi.org/10.5075, 156.

Toit, C.D., and D.B. Davidson. 1995. Wiregrid: A NEC2 Pre-processor. *Journal of Applied Computational Electromagnetics Society* 10: 31–39.

Chapter 2
Theoretical Background for the Computation of Radar Cross-Section (RCS) Using NEC

The entire theoretical framework of NEC starting from the definition of integral equations to the solution of matrix equation is briefly introduced in this chapter. The basis functions used for representing the unknown current densities and the weight functions used for testing are also described in detail. Further, the strategy used for parallelization of NEC (Rubinstein et al. 2003) is also included here.

2.1 Formulation of Integral Equations (IE)

NEC employs both an electric-field integral equation (EF-IE) and a magnetic-field integral equation (MF-IE) for evaluating the electromagnetic behaviour of conducting bodies under consideration. EF-IEs are mainly used in the case of thin-wire geometries with small conductor volumes. On the other hand, MF-IE, which cannot be applied for the thin-wire case, is more suitable for voluminous structures with smooth surfaces. EF-IE and MF-IE are coupled together for a structure containing both wires and surfaces. The theoretical formulation presented in the following sections is based on the work reported in (Burke and Poggio 1981).

2.1.1 Electric-Field Integral Equation (EF-IE)

The integral representation for the electric field due to a surface current distribution $\vec{J}_S$ on a perfectly conducting body (S) is given by,

$$\vec{E}(\vec{r}) = (-j\eta/4\pi k) \int_S \left(\vec{J}_S(\vec{r'}) \cdot \bar{\bar{G}}(\vec{r}, \vec{r'}) \right) dA' \tag{2.1}$$

© The Author(s), under exclusive license to Springer Nature Singapore Pte Ltd. 2021
V. Joy et al., *Fundamentals of RCS Prediction Methodology using Parallelized Numerical Electromagnetics Code (NEC) and Finite Element Pre-processor*,
SpringerBriefs in Computational Electromagnetics,
https://doi.org/10.1007/978-981-15-7164-0_2

where,

$$\overline{\overline{G}}\left(\vec{r}, \vec{r}'\right) = \left\{k^2\overline{\overline{I}} + \nabla\nabla\right\}g\left(\vec{r}, \vec{r}'\right)$$

$$g\left(\vec{r}, \vec{r}'\right) = \frac{e^{-jk|\vec{r}-\vec{r}'|}}{|\vec{r} - \vec{r}'|}$$

$$k = \omega\sqrt{\mu_0\varepsilon_0}; \quad \eta = \sqrt{\frac{\mu_0}{\varepsilon_0}}; \quad \overline{\overline{I}} = \hat{x}\hat{x} + \hat{y}\hat{y} + \hat{z}\hat{z}$$

$\vec{r}$ and $\vec{r}'$ represent observation point and source point, respectively.

The integral equation for the current induced on S by an incident field $\vec{E}^{\text{Inc}}$ is obtained by imposing the following boundary condition for $\vec{r} \in S$:

$$\hat{n}(\vec{r}) \times \left[\vec{E}^{\text{Sca}}(\vec{r}) + \vec{E}^{\text{Inc}}(\vec{r})\right] = 0 \tag{2.2}$$

where $\hat{n}(\vec{r})$ is the unit normal vector to the surface at $\vec{r}$ and $\vec{E}^{\text{Sca}}$ is the scattered field due to the induced current $\vec{J}_s$. Substituting (2.1) for $\vec{E}^{\text{Sca}}$ in (2.2) yields the final EF-IE as,

$$-\hat{n}(\vec{r}) \times \vec{E}^{\text{Inc}}(\vec{r}) = (-j\eta/4\pi k) \times \hat{n}(\vec{r}) \times \oint_S \vec{J}_s(\vec{r}') \cdot \left(k^2\overline{\overline{I}} + \nabla\nabla\right)g\left(\vec{r}, \vec{r}'\right)\mathrm{d}A'$$

$$\tag{2.3}$$

Since NEC employs thin cylindrical wires for representing the conducting surface (S), the vector integral equation in (2.3) will be transformed to a scalar integral equation. The assumptions applicable for the thin-wire approximation used in most of the cases are given below:

(i)   Currents in the transverse direction can be ignored with respect to the currents along axial direction of the wire.
(ii) Circumferential variation in the currents along axial direction can be ignored.
(iii) The unknown current may be expressed as a filament along the wire axis.
(iv) The boundary condition pertaining to the electric field is required to be imposed only in the axial direction.

2.1.2 Magnetic-Field Integral Equation (MF-IE)

The integral representation for the magnetic field due to a surface current distribution $\vec{J}_s$ on a perfectly conducting body is given by,

$$\vec{H}(\vec{r}) = (1/4\pi) \int_S \left(\vec{J}_s(r') \cdot \nabla' g(\vec{r}, \vec{r}') \right) dA' \tag{2.4}$$

where the differentiation is with respect to the integration variable $\vec{r}'$.

If an external incident field $\vec{H}^{\text{Inc}}$ induces a current $\vec{J}_s$ on S, then the total magnetic field within the perfectly conducting body vanishes. Hence, the boundary condition for $\vec{r}$ immediately within the surface S can be expressed as,

$$\left[\vec{H}^{\text{Sca}}(\vec{r}) + \vec{H}^{\text{Inc}}(\vec{r}) \right] = 0 \tag{2.5}$$

where $\vec{H}^{\text{Inc}}$ is the incident field without the structure and $\vec{H}^{\text{Sca}}$ is the scattered field due to the induced current $\vec{J}_s$ given by (2.4).

Substituting $\vec{H}^{\text{Sca}}$ in (2.5) yields the final MF-IE as,

$$-\hat{n}(\vec{r}_0) \times \vec{H}^{\text{Inc}}(\vec{r}_0) = \hat{n}(\vec{r}_0) \times (1/4\pi) \lim_{\vec{r} \to \vec{r}_0} \int_S \left(\vec{J}_s(\vec{r}') \times \nabla' g(\vec{r}, \vec{r}') \right) dA' \tag{2.6}$$

where $\hat{n}(\vec{r}_0)$ is the normal vector directed outwards at the surface point $\vec{r}_0$.

Similar to EF-IE, the vector integral Eq. (2.6) can also be resolved into two scalar expressions in the directions of the orthogonal surface vectors $\hat{t}_1$ and $\hat{t}_2$ where

$$\left[\hat{t}_1(\vec{r}_0) \times \hat{t}_2(\vec{r}_0) \right] = \hat{n}(\vec{r}_0) \tag{2.7}$$

2.1.3   EF-IE/MF-IE Combined Equation

The EF-IE/MF-IE combined equation is used for a structure consisting of both wires and surfaces. In such cases, $\vec{r}$ in (2.5) is restricted to wires and the integral for $\vec{E}^{\text{Sca}}(\vec{r})$ extends over the complete structure. Similarly, $\vec{r}_0$ in (2.6) is restricted to surfaces and the integrals for $\vec{H}^{\text{Sca}}(\vec{r})$ extends over the entire structure. The integral can be simplified by using the thin-wire approximation on wires.

The final coupled integral equations (Burke and Poggio 1981) are,

$$-\hat{s} \cdot \vec{E}^{\text{Inc}}(\vec{r}) = (-j\eta/4\pi k) \int_L I(s') \left[\left(k^2 \, \hat{s} \cdot \hat{s}' \right) - \left(\frac{\partial^2}{\partial s \, \partial s'} \right) \right] g(\vec{r}, \vec{r}') dS'$$

$$- (j\eta/4\pi k) \int_{S_1} \vec{J}_s(\vec{r}') \left[\left(k^2 \, \hat{s} \right) - \left(\nabla' \frac{\partial}{\partial s} \right) \right] g(\vec{r}, \vec{r}') dA';$$

for $\vec{r}$ on surfaces of wire

$$\tag{2.8}$$

$$\hat{t}_2(\vec{r}) \cdot \vec{H}^{\text{Inc}}(\vec{r}) = (-1/4\pi)\hat{t}_2(\vec{r}) \cdot \int_L I(s')\left[\hat{s}' \times \nabla' g\left(\vec{r}, \vec{r}'\right)\right] dS'$$

$$- \left[(1/2)\hat{t}_1(\vec{r}) \cdot J_S(\vec{r})\right] - (1/4\pi)\int_{S_1} \hat{t}_2(\vec{r})$$

$$\cdot \left[\vec{J}_s(\vec{r}') \times \nabla' g\left(\vec{r}, \vec{r}'\right)\right] dA';$$

for $\vec{r}$ on surfaces excluding wires   (2.9)

$$-\hat{t}_1(\vec{r}) \cdot \vec{H}^{\text{Inc}}(\vec{r}) = (1/4\pi)\hat{t}_1(\vec{r}) \cdot \int_L I(s')\left[\hat{s}' \times \nabla' g\left(\vec{r}, \vec{r}'\right)\right] dS'$$

$$- \left[(1/2)\hat{t}_2(\vec{r}) \cdot J_S(\vec{r})\right] + (1/4\pi)\int_{S_1} \hat{t}_1(\vec{r})$$

$$\cdot \left[\vec{J}_S(\vec{r}') \times \nabla' g\left(\vec{r}, \vec{r}'\right)\right] dA';$$

for $\vec{r}$ on surfaces excluding wires   (2.10)

where s is the parameter corresponding to distance along the wire axis at $\vec{r}$ and $\hat{s}$ is the unit vector tangent to the wire axis at $\vec{r}$. $\int_L$ and $\int_{S_1}$ represents integration over wires and integration over surfaces excluding wires, respectively.

2.2 Computation of Scattered Fields

The scattered fields due to induced currents can be computed using a far-field approximation. This is applicable when the observation points are far apart from the induced current distribution by distances much larger compared to the wavelength as well as the dimensions of the current distribution. The far-zone scattered fields from a conducting body consisting of a wire region with contour L having linear current distribution $\vec{I}(s)$ and a surface region S with surface current density $\vec{J}_s(\vec{r})$ are given by,

$$\vec{E}(\vec{r}_0) = (jk\eta/4\pi) \times (e^{-jkr_o}/r_o) \times \left\{ \int_L \left[(\hat{k} \cdot \vec{I}(s))\hat{k} - \vec{I}(s)\right] e^{jk \cdot \vec{r}} \right.$$

$$\left. + \int_S \left[(\hat{k} \cdot \vec{J}_s(\vec{r}))\hat{k} - \vec{J}_s(\vec{r})\right] e^{jk \cdot \vec{r}} dA \right\}$$ (2.11)

where $\hat{k} = \frac{\vec{r}_o}{|\vec{r}_o|}$, $k = \frac{2\pi}{\lambda}$ and $\vec{r}_o$ is the position of the observation point.

The contour integral can be computed in closed form over each wire segment for the cosine, sine, and constant components of the basis functions and this in turn becomes a summation over the wire segments. Since the surface current on the patch is represented by a delta function at the centre of the patch, the second integral transforms to a summation over the patch elements.

For an incident plane wave with electric field $\vec{E}^{\text{Inc}}$, the NEC program constants are defined such that the radar cross-section (RCS) σ/λ^2 is displayed under the column gain in the output file.

$$\frac{\sigma}{\lambda^2} = 4\pi \frac{\left|\vec{E}^{\text{Sca}}\right|^2}{\left|\vec{E}^{\text{Inc}}\right|^2} \tag{2.12}$$

where $\vec{E}^{\text{Sca}}$ is given by (2.11).

2.3 Numerical Solution of Integral Equations

The integral equations described in the previous sections are solved numerically in NEC by using the method of moments (MoM). Method of moments starts with a conventional linear-operator equation as,

$$L(f) = u \tag{2.13}$$

where f is an unknown response, u is a known excitation and L is an integral operator.

The formulation proceeds by expressing the unknown function f as a sum of known basis functions f_j (defined in the domain of the integral operator L) as,

$$f = \sum_{j=1}^{N} \beta_j f_j \tag{2.14}$$

The unknown coefficients (β_j) are obtained by taking the inner product of (2.13) with a set of weighting functions w_i, i.e.

$$\langle w_i, L(f)\rangle = \langle w_i, u\rangle; \quad i = 1, 2, \ldots, N \tag{2.15}$$

The inner product in the present case can be defined as,

$$\langle p, q\rangle = \int_S (p(\vec{r}) q(\vec{r})) dA$$

where integration is over the surface of the structure.

Substituting (2.14) in (2.15) gives

$$\sum_{j=1}^{N} \beta_j \langle w_i, L(f_j) \rangle = \langle w_i, u \rangle; \quad i = 1, 2, \ldots, N \tag{2.16}$$

The above equation can be expressed in matrix form as,

$$[G][B] = [U] \tag{2.17}$$

where $G_{ij} = \langle w_i, L(f_j) \rangle$; $B_j = \beta_j$; $U_i = \langle w_i, u \rangle$.

The solution to (2.17) is

$$[B] = [G]^{-1}[U] \tag{2.18}$$

2.3.1   Selection of Basis Functions and Weight Functions

The critical factors determining the accuracy and convergence of MoM are the weight functions and the basis functions. NEC employs different basis and weight functions unlike Galerkin's method. The weighing functions (w_i) are chosen as a set of delta functions in accordance with the collocation method, where the integral equation is enforced at specific sample points on the geometry, i.e.

$$w_i(\vec{r}) = \delta(\vec{r} - \vec{r}_i)$$

where $\{\vec{r}_i\}$ are the set of points on the conducting surface. In NEC, conducting wires are sub-divided into short linear segments where a sample point is defined at the centre of each segment. On the other hand, surfaces are represented by a set of flat patches with a sample point defined at the patch centre.

The linear current density on wires and surface current density on surfaces are the unknown quantities in NEC which are to be represented by a set of known basis functions. NEC uses sub-domain basis functions to simplify the calculation of inner product integrals and also to ensure that the matrix G is well conditioned.

In the case of wires, the current on each segment is approximated as a sum of three terms, viz., a sine, a cosine, and a constant. The total current on ith segment can thus be expressed as,

$$I_i(s) = A_i \sin k(s - s_i) + B_i \cos k(s - s_i) + C_i, \quad |s - s_i| < \frac{\Delta_i}{2} \tag{2.19}$$

where s_i is the value of s at the centre of segment i, Δ_i is the length of segment i, and k is the free space wave number.

The additional equations required to evaluate all the unknown coefficients in (2.19) for each and every segment will be provided by the local boundary conditions. The continuity of current and charge is one such condition to be enforced at a junction of two segments with uniform radius. In the case of a junction with more than two segments of different radii, the continuity of current may be generalized to Kirchhoff's current law.

The final solution requires the computation of the electric field at each segment due to these currents. NEC uses three approximations of the integral equation kernel for the computation of fields. These include (i) thin-wire approximation for most of the cases, (ii) extended thin-wire approximation for thick wires, and (iii) current element approximation for the case where segments are separated far apart by a considerable distance. In all the three methods used for the evaluation of field at a particular segment due to the induced current on a different segment, the field computation is performed on the surface of the observation segment.

NEC models surfaces using small flat patches and the unknown surface current on each patch are expressed as a summation of a set of pulse functions except where a wire connection is present. When a wire connects to a surface, a special local boundary condition has to be enforced as explained in (Burke and Poggio 1981).

The unknown surface current density over M_p patches can be written as,

$$\overrightarrow{J}_s(\vec{r}) = \sum_{i=1}^{M_p} (J_{1i}\hat{t}_{1i} + J_{2i}\hat{t}_{2i})v_i(\vec{r}) \tag{2.20}$$

where $\hat{t}_{1i} = \hat{t}_1(\vec{r}_i)$, $\hat{t}_{2i} = \hat{t}_2(\vec{r}_i)$, $\vec{r}_i$ is the position of the centre of patch i, $v_i(\vec{r}) = 1$ for $\vec{r}$ on patch i and 0 otherwise. Thus, every patch has two unknown quantities and the equation is enforced for each vector component at the sample point in a given patch.

2.3.2 Solution of Matrix Equation

The matrix equation obtained by appropriately substituting the basis functions and weight functions in (2.16) has to be finally solved for the unknown currents. The equation can be alternately expressed as,

$$[G][I] = [E] \tag{2.21}$$

where $[G]$ is the interaction matrix, $[I]$ is the column vector with unknown currents, and $[E]$ is the excitation vector. NEC employs Gauss elimination method for solving (2.21). Here, the matrix G is first factorized into the product of upper and lower triangular matrices, i.e.

$$[G] = [L][U]$$ (2.22)

The matrix equation then becomes,

$$[L][U][I] = [E]$$ (2.23)

Let,

$$[L][F] = [E]$$ (2.24)

where,

$$[U][I] = [F]$$ (2.25)

Equation (2.24) is first solved for F by forward substitution. Then (2.25) is solved for I by backward substitution. The order of the matrix Eq. (2.21) for a structure having N_w wires segments and N_p patches is given by,

$$N = N_p + N_w$$

The final matrix Eq. (2.21) can be written in terms of sub-matrices as,

$$\begin{bmatrix} P & Q \\ R & S \end{bmatrix} \begin{bmatrix} I_W \\ I_P \end{bmatrix} = \begin{bmatrix} E_W \\ H_P \end{bmatrix}$$ (2.26)

where I_w is the column vector of segment basis function amplitudes, I_p is the column vector of patch-current amplitudes, E_w is the LHS of (2.8) evaluated at segment centres, and H_p is the LHS of (2.9) and (2.10) evaluated at patch centres.

The sub-matrix element P_{ij} denotes the electric field at the centre of segment i due to the basis function centred on segment j. On the other hand, the sub-matrix element S_{ij} denotes the tangential magnetic-field component at patch u due to the surface current pulse on patch v where

$$u = 1 + \text{Int}\left[\frac{i-1}{2}\right]; \text{ Int[ ]indicates truncation}$$

$$v = 1 + \text{Int}\left[\frac{j-1}{2}\right]$$

The elements in sub-matrix Q denote the electric fields due to surface current pulses. On the other hand, the elements in R denote magnetic fields due to segment basis functions.

2.4 Strategy for Parallelization

The structure of NEC can be broadly divided into two parts, namely the input section and the calculation section. The input section reads in the geometrical data pertaining to the structure under consideration as well as the program control commands. The calculation section on the other hand computes the elements of the interaction matrix [G] which holds the geometrical and electromagnetic information about the model. This section also deals with the solution of (2.21) by Gauss elimination method. This final step is carried out in two routines. The first one performs the computationally intensive task of LU decomposition of [G]. Later, it is used along with the excitation vector [E] in another routine to obtain the solution. The computation of the constituent elements of [G] and evaluation of the final solution to the matrix equation accounts for more than 90% of the computation time (Burke and Poggio 1981). These parts of the code need to be parallelized for the same reason.

The parallelized version of NEC used in this book is in accordance with (Rubinstein et al. 2003) where the subroutines responsible for LU decomposition (FACTR) and solution of matrix equation (SOLVE) have been revised to work in a parallel environment (Rubinstein 2004). This version distributes the interaction matrix among the available processors and the matrix elements are locally and individually calculated by their respective processors. The final system of linear equations is also solved using a parallel version of the Gauss-Doolittle algorithm. In this way, the combined processing power and memory of several processors working together as a cluster can be efficiently utilized. The code can automatically detect the number of processors and hence exploit the available memory and computing resources. The parallelized code (Rubinstein et al. 2003) is designed to run on distributed memory parallel supercomputers and it can work on any system/cluster supporting conventional message passing parallel environments such as Message Passing Interface (MPI) and Parallel Virtual Machine (PVM). Since it is completely based on the original NEC code, the input files need not be changed and the output format remains the same too.

The efficient distribution of the information stored in the matrices among the available processors is very much critical for the successful parallelization of LU decomposition and the parallel solution of the system of equations. For this reason, the interaction matrix and the excitation vector have to be divided into smaller submatrices which will be local to the respective processors. In (Rubinstein et al. 2003), two-dimensional block-cyclic distribution has been used to uniformly distribute the computation effort among all available processors and it is discussed in the following paragraphs.

To illustrate two-dimensional block-cyclic distribution, consider a parallel computer with four processors numbered as 1, 2, 3, and 4, as shown in Fig. 2.1. A matrix is to be distributed among them. In a conventional way, as shown in Fig. 2.1a, the matrix is distributed column wise to each processor. For ease of understanding, the contents of processor 1 are shaded blue. Here, once all the elements that correspond to processor 1 are assigned, it will become idle for the rest of the time. The

Fig. 2.1 a 1D column
distribution, **b** 2D
block-cyclic distribution

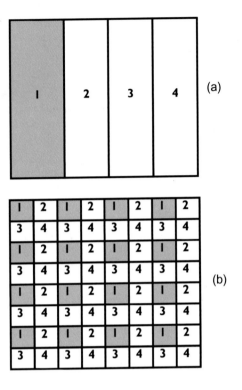

same situation repeats for the rest of the processors too. On the other hand, two-dimensional block-cyclic distribution as shown in Fig. 2.1b guarantees that, on an average, all the processors remain active from the start to the end of the algorithm. A square processor grid is a desirable attribute for better efficiency as it maintains equitable distribution of the computational load. But, in a practical scenario, the number of processors varies from one machine to another. Therefore, the squarest possible processor grid on the running platform is automatically calculated at the beginning of the algorithm. ScaLAPACK library has been used in (Rubinstein et al. 2003) for implementing the block-cyclic decomposition in NEC. A brief description of ScaLAPACK and MPI is given in Appendix A.

It is noted that the formation of the interaction matrix on a single processor and then the distribution among other processors is not an efficient solution for memory problem. The interaction matrix computation is therefore parallelized in such a manner so as to produce an already distributed matrix as it is being computed. Each processor on the grid will have adequate information required to determine the elements of [G] once the input section is completed. An additional function is employed to notify each processor whether the element (i, j) being calculated corresponds to its local sub-matrix. In that case, another sub-routine will compute the local values of (i, j) for the same element so as to be assigned to the corresponding local matrix. Once the matrix generation is completed, each processor will hold its local version of the interaction matrix and there would not be any further need for communication among the processors. Once the matrix equation is solved, data is compiled by a block-cyclic composition routine and all the processors become aware of the solution.

An approximation of the memory requirements in bytes at a local node for the parallelized model of NEC is given by (Rubinstein et al. 2003),

$$\text{Memory} = \frac{\left(16 \times N_{seg}^2\right)}{P}$$

where P is the number of processors being used during the execution of the program and N_{seg} is the number of segments. On the other hand, the original version of NEC which is not parallelized requires $(16 \times N_{seg}^2)$ bytes of memory. Figure 2.2 presents the difference in memory requirements at a particular node between the original version of NEC and its parallelized counterpart.

Considering a structure meshed with a particular number of segments, the variation in the run time of NEC with respect to the number of processors has been studied in (Rubinstein et al. 2003), keeping all other parameters fixed. When the number of processors was increased from one to four, it has been found that the total run time reduced by almost 98%, thereby emphasizing the efficiency of parallelization. However, the number of processors has to be judiciously chosen based on the electrical dimensions of the structure under consideration, as employing large number of processors may also increase the run time due to communication delay.

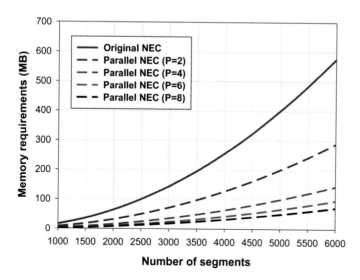

Fig. 2.2 Variation in memory requirements of NEC at a particular node with respect to number of segments

References

Burke, G.J., and A.J. Poggio. 1981, January. Numerical Electromagnetics Code (NEC)—Method of Moments, Lawrence Livermore National Laboratory, Livermore, California, Technical Document, Rep. UCID-18834, 719.

Rubinstein, A. 2004, January. Simulation of Electrically Large Structures in EMC Studies: Application to Automotive EMC, Ph.D. dissertation, École polytechnique fédérale de Lausanne (EPFL), https://doi.org/10.5075, 156.

Rubinstein, A., F. Rachidi, M. Rubinstein, and B. Reusser. 2003. A parallel implementation of NEC for the analysis of large structures. *IEEE Transactions on Electromagnetic Compatibility* 45 (2): 177–188.

Chapter 3
Modelling Guidelines and Input/Output Formats in NEC

The guidelines to be followed while modelling geometries for electromagnetic analysis using NEC are presented in this chapter. NEC uses short, straight segments for modeling wires and flat patches for modeling surfaces (Burke and Poggio 1981). All the geometries under consideration must be therefore modeled with a series of segments along wires and with patches covering surfaces. The structure of NEC input and output files is also discussed in this chapter.

3.1 Modelling of Wires

The basic parameters defining a wire segment are the coordinates of its two end points and its radius. The geometry of a wire segment oriented along Z-axis is shown in Fig. 3.1. Both geometrical and electrical considerations are involved in the wire grid modelling of a structure with segments. From the geometrical aspect, the segments are supposed to closely follow the paths of conductors, using a piece-wise linear fit on curves (Burke and Poggio 1981). One of the critical factors from electrical point of view is the segment length (Δ) compared to the wavelength (λ). The general rule of thumb for segment length can be expressed as,

$$\Delta \ll \frac{\lambda}{10}$$

where λ is the wavelength at desired frequency.

Slightly lengthier segments may be used for long wires with no abrupt variations while shorter segments ($\Delta < \lambda/20$) may be used for modelling important regions. Since current is calculated at the center of each segment, the segment length determines the resolution in solving for the unknown current. Shorter segments ($\Delta < \lambda/1000$) should be avoided to eliminate numerical inaccuracy.

© The Author(s), under exclusive license to Springer Nature Singapore Pte Ltd. 2021
V. Joy et al., *Fundamentals of RCS Prediction Methodology using Parallelized Numerical Electromagnetics Code (NEC) and Finite Element Pre-processor*,
SpringerBriefs in Computational Electromagnetics,
https://doi.org/10.1007/978-981-15-7164-0_3

Fig. 3.1 Geometry of a wire
segment

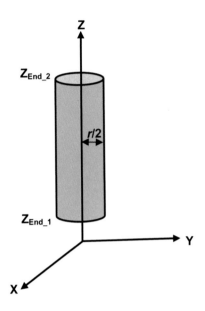

Another important electrical parameter is the wire radius (r) compared to wavelength (λ). This is limited by the approximations (thin-wire kernel or extended thin-wire kernel) used in the evaluation of EFIE. In both of the approximations, circumferential variation of current is not permitted and only axial currents on a segment are considered. Computational studies have established that with the thin-wire kernel, Δ/r, greater than 8 is required for errors less than 1%, and for the extended thin-wire kernel, it can be as small as 2 for achieving similar accuracy. It is advisable to use the extended thin-wire kernel option [by including EK command, (Burke and Poggio 1981)] whenever a model includes segments with Δ/r less than about 2. Extended thin-wire kernel option is typically used at free wire ends and between parallel, connected segments. However, bends in wires always have to be treated with thin-wire kernel approximation. This in turn implies that segments with small Δ/r are to be avoided at bends. The reason is that in such a case, the center of one segment falls within the radius of another segment leading to erroneous results.

Electrical connectivity between segments is another aspect to be taken care of, for ensuring the continuity of current and charge density along wires, at junctions, and at wire ends. NEC requires the electrically connected segments to have coincident end points for the satisfactory compliance with these conditions. The code does not allow current flow if segments intersect other than at their ends. Therefore, it is always advisable to use identical coordinates for connected segment ends. If the wire ends are separated by distances less than 10^{-3} times the length of the shortest segment, then the segments will be treated as connected.

Other important modelling guidelines are listed below (Burke and Poggio 1981):

- The angle of intersection of wire segments should be large enough to prevent overlaps.

- Due to a dimension limitation in the code, the number of wires joined at a single junction cannot exceed thirty.
- It is always advisable to keep the adjacent wires several radii apart. The segments which are parallel and close together need to be properly aligned to avoid incorrect current perturbation.
- Overlapping between segments is to be strictly avoided. The reason is that the division of current between two overlapping segments is indeterminate.
- The huge change in radius between connected segments should not happen abruptly. It should be done in steps over several segments.

3.2   Modelling of Surfaces

Surface patches are used for modelling closed conducting surfaces, just like segments are used to model wires. The patches should be efficiently laid out so as to completely cover the surface to be modelled and should also conform closely to curved surfaces (Burke and Poggio 1981). The basic parameters used to define a surface patch are:

- Cartesian coordinates of the center of the patch (x, y, z).
- Components of the unit normal vector directed outwards $(\hat{n})$.
- Area of the patch (Ar).

The geometry of a surface patch along with several options for patch shape is shown in Fig. 3.2. $\hat{t}_1$ and $\hat{t}_2$ (given by $\hat{n} \times \hat{t}_1$) are orthogonal surface vectors along which the surface currents will be computed for each patch.

The patch can be of rectangular, triangular, quadrilateral, or arbitrary shape. Since NEC does not integrate over patches, except at wire connection, the shape of the patch does not affect the final output. When a wire connects to a surface, the wire end as well as the patch center should have identical coordinates. This situation is treated differently in NEC. The patch size measured in wavelengths is critical for the accuracy of results similar to wire modelling. The general rule of thumb insists a minimum of about 25 patches per λ^2 of surface area with the maximum dimension of an individual patch not exceeding $0.04 \lambda^2$. Similar to wire grid modelling, large patches may be used for modelling large smooth surfaces and smaller ones for critical areas. Detailed discussions on surface modelling can be found in (Burke and Poggio 1981).

3.3   Format of NEC Input File

The NEC input file consists of a series of command lines carrying information regarding structure geometry and program execution. The type of command line is identified at the beginning of it using a two-letter alphabetic code and the command lines having numeric data are written with integer numbers first followed by real

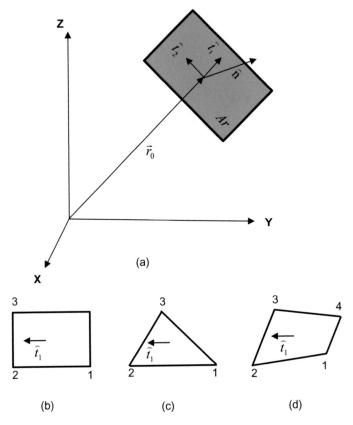

Fig. 3.2 Surface patch **a** general geometry **b** rectangular patch **c** triangular patch **d** quadrilateral patch

numbers. The command lines in NEC can be classified into three types as given below (Burke and Poggio 1981):

- *Comment lines*: They contain a brief description of the problem being dealt with and the data on these lines will be printed at the beginning of the output file. The NEC input file begins with one or more comment lines and it can have any alphabetic and numeric characters. They begin with a two-letter identifier (CM). A CE command line is used to notify the end of comment lines. The line immediately after a CE line should be a geometry command line.
- *Geometry commands*: They are used to describe the geometry of the structure under consideration. For instance, a straight wire with an arbitrary number of segments can be defined by a single input line specifying the coordinates of the end points and the number of segments. Several other commands are also available for performing various operations like translation, reflection, rotation, etc. The user can also assign tag numbers to segments for referring to them later.

Table 3.1 List of geometry commands

Geometry commands	Purpose
GA	To specify a wire arc
GE	To signal end of geometry data
GF	To use numerical Green's function
GM	To shift and duplicate structure
GS	To scale structure dimensions
GW	To specify wire with a number of segments
GX	To reflect structure (symmetry)
GR	To generate cylindrical structure (symmetry)
SP	To specify surface patch
SM	To generate multiple surfaces patches

The geometry command lines have two fields (three columns each) for integer numbers followed by real-number fields (ten columns each). The list of important geometry commands is given in Table 3.1.

- *Program control commands*: They carry information regarding the electrical parameters of the model and they also request for data computations. The program control commands have four integer fields followed by real-number fields. They immediately follow the structure geometry commands. The list of program control commands is given in Table 3.2.

In the case of program control commands, there is no fixed order in which the commands are to be specified. The required electrical parameters are initialized first followed by requests for computation of near fields, radiated fields, and currents. Default values are assigned to those parameters which are not mentioned in the input data with an exception for the excitation (EX) command which must be set. The currents and requested quantities are calculated at the first occurrence of RP, NE, and NH commands. Subsequent occurrences of these commands trigger field computation using the previously computed currents. The near-field and radiation pattern requests can be repeated any number of times in the input file but with an exception when multiple frequencies are mentioned on a single FR command. Here, only a single NE or NH command and a single RP command will be applicable for each and every frequency. The format of frequently used geometry commands and program control commands is given in Appendix B.

3.4 Format of NEC Output File

The output file starts with a set of comments on the problem followed by the geometrical information and computational results. The section titled "structure specification" lists the geometry command lines along with the relevant information. The total

Table 3.2 List of program control commands

Program control command	Purpose
EK	To specify the application of extended thin-wire kernel
FR	To specify operating frequency
GN	To specify ground parameters
KH	To specify interaction approximation range
LD	To specify structure impedance loading
EX	To specify structure excitations
NT	To specify two-port networks
TL	To specify transmission line
CP	To specify coupling calculations
EN	To specify the end of data flag
GD	To specify additional ground parameters
NE	To request the calculation of near electric fields
NH	To request the calculation of near magnetic fields
NX	To specify next structure flag
PQ	To request the printing of wire charge density
PT	To request the printing of wire current
RP	To request the calculation of radiation pattern
WG	To write numerical Green's function file
XQ	To initiate program execution

number of segments and patches used in the model is printed once the GE command is read. Furthermore, all junctions with three or more wire connections are listed under the section titled "multiple wire junctions" and all the parameters pertaining to individual segments like coordinates of segment center, segment length, orientation angles, wire radius, connection data, and tag number are summarized under "segmentation data". If patches are used in the model, components of the unit normal vector, coordinates of the center of the patch and patch area will be displayed under the heading "surface patch data". All the lines immediately following the geometry commands are displayed exactly as they are read by the code. When a request for radiated fields is made, fields as well as currents are printed, and if a plane wave excitation is used, the gain columns will contain the bi-static radar cross-section. The format of a typical NEC output file is given in Appendix C.

Reference

Burke, G.J., and A.J. Poggio. 1981 January. Numerical Electromagnetics Code (NEC)—Method of Moments, Lawrence Livermore National Laboratory, Livermore, California, Technical Document, Rep. UCID-18834, 719.

Chapter 4
Methodology for the Computation of RCS using Parallelized NEC and Finite Element Pre-processor

The prediction of RCS of potential structures (Bowman et al. 1969; Sohel and Rozas 1974) basically involves the computation of reradiated fields by the currents induced on the surface by incident electromagnetic waves. In this context, it is noted that NEC is based on the numerical solution of integral equations for these induced currents. As far as conducting surfaces are considered, it is difficult to surpass the accuracy and efficiency of NEC (Uluisik et al. 2008; Sevgi 2003). However, one of the most cumbersome steps in NEC is the translation of geometry data into a format compatible with NEC modelling guidelines. It is an extremely error-prone process, and furthermore, the code will often run in the presence of such errors and produce results that are conceivable but incorrect. Hence, an appropriate pre-processor is an essential prerequisite for the accurate prediction of RCS using NEC. Altair HyperMesh, a market-leading, multidisciplinary finite element pre-processor, can be a desirable choice for this purpose.

4.1 Overview of Finite Element Pre-processor

Altair HyperMesh is an efficient finite element pre-processor that provides superior quality meshing coupled with a user-friendly visual environment. It has got direct interfaces to most of the commercial computer-aided design (CAD) and computer-aided engineering (CAE) systems. Furthermore, it offers a broad range of user-friendly tools to model even the most complex structural geometries. An overview of the graphical user interface (GUI) of HyperMesh is given in Fig. 4.1.

© The Author(s), under exclusive license to Springer Nature Singapore Pte Ltd. 2021
V. Joy et al., *Fundamentals of RCS Prediction Methodology using Parallelized Numerical Electromagnetics Code (NEC) and Finite Element Pre-processor*,
SpringerBriefs in Computational Electromagnetics,
https://doi.org/10.1007/978-981-15-7164-0_4

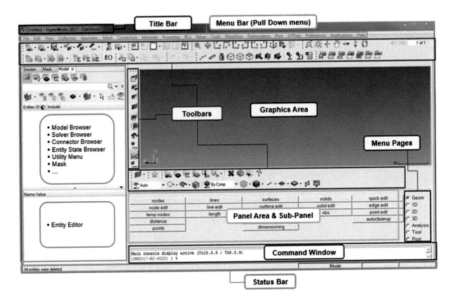

Fig. 4.1 Graphical user interface (GUI) of HyperMesh

The robust surface meshing module in HyperMesh has the capability to create and optimize a particular mesh based on user specifications. It also supports a solid meshing module which swiftly produces high-quality meshes for multiple volumes and has several standard and advanced options to connect, separate, or split solid geometries for tetrahedral-based or hexahedral-based meshing. Segregation of these models can be done quickly and easily with the powerful visualization features of HyperMesh.

HyperMesh is compatible with all popular CAD file formats like ACIS, IGES, CATIA V4/V5/V6, Intergraph, Inspire, JT, PTC Creo, Parasolid, SolidWorks, Tribon, STEP, NX, etc. It also provides direct support (import/export) to popular solvers like ANSYS, Abaqus, Nastran, OptiStruct, RADIOSS, etc. HyperMesh also has efficient tools to clean up imported geometries that contain surfaces with various defects like overlaps, gaps, and misalignments. The elimination of these defects and suppression of the shared boundaries between adjacent surfaces helps in meshing across larger and more important regions of the geometry. This in turn increases meshing quality and speed.

Another powerful utility available in HyperMesh is the BatchMesher option which automatically generates superior meshes for large assemblies, with minimal manual meshing tasks. BatchMesher also allows the user to specify meshing criteria, geometry clean-up parameters, and the output file format.

4.2 Revised Methodology for the Computation of RCS using Parallelized NEC and Finite Element Pre-processor

The revised methodology for the computation of RCS using NEC in conjunction with HyperMesh is shown in Fig. 4.2. The herculean task of generating wire-grid-based mesh data catering to NEC modelling guidelines can be accurately performed in HyperMesh. The inclusion of HyperMesh has therefore significantly simplified the meshing procedure.

The steps to be followed for the wire-grid-based modelling of geometries in HyperMesh are given below:

Step 1: Modelling of geometry/Importing of CAD model in HyperMesh

The geometries can either be modelled directly in HyperMesh, or the CAD model can be imported in suitable format. The tolerance values for geometry clean-up and meshing have to be appropriately set thereafter. HyperMesh initially displays a list of supported user profiles, and the required one has to be selected from this list. Nastran profile is chosen in this book. Based on the user profile chosen, output format will vary and the extraction algorithm has to be appropriately framed.

Step 2: Automeshing of the model

Once the geometry is appropriately modelled/imported, the automesh option has to be used for segmentation of the surface using suitable mesh elements (triangles, quadri-laterals, etc.). The element size to be specified in HyperMesh has to be calculated in accordance with the criterion mentioned in Chap. 3.

Step 3: Wire-grid-based meshing of the model

In this step, the detach option has to be used for separating the mesh elements from the structure. Once this is done, the detached edges have to be identified and a suitable property has to be assigned to the edge elements. The required wire radius in accordance with NEC modelling guidelines can be mentioned while defining the

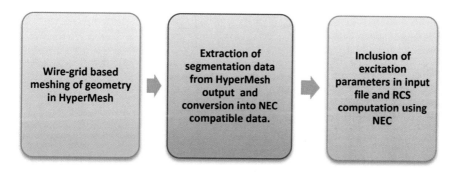

Fig. 4.2 Workflow for the computation of RCS using NEC in conjunction with HyperMesh

property. Finally, the line mesh option has to be used for assigning wire elements of appropriate radius along the identified edges. Furthermore, similar mesh elements appearing more than once have to be eliminated by using duplicate option.

Step 4: Processing of HyperMesh output

The output file can be generated in any suitable format. Nastran file format is used in this book. The format of a typical Nastran file is given in Appendix C. The mesh data has to be then converted into NEC format using a suitable format conversion algorithm depending on the user profile chosen. The mesh data in NEC format has to be then incorporated into the NEC input file with the required program command lines. The details regarding the required libraries like ScaLAPACK, BLAS, etc., and the procedure for the compilation and execution of the parallelized version of NEC are given in Appendix A. The flow chart summarizing the entire workflow for the computation of RCS using NEC in conjunction with HyperMesh is presented in Fig. 4.3.

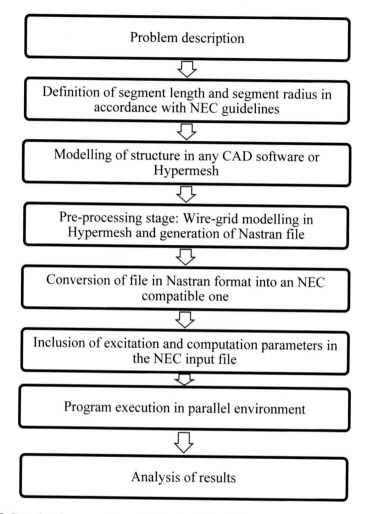

Fig. 4.3 Procedure for computation of RCS using NEC and HyperMesh

References

Bowman, J.J., T.B.A. Senior, and P.L.E. Uslenghi. 1969. *Electromagnetic and Acoustic Scattering by Simple Shapes*, 380–413. Amsterdam: North-Holland, ISBN:13 978-0-891-16885-0.

Sevgi, L. 2003. *Complex Electromagnetic Problems and Numeric Simulation Approaches*, 179–218. Wiley-IEEE Press. ISBN 978-0-471-43062-9.

Sohel, M.S., and P. Rozas. 1974. *Radar Cross Sections of Standard and Complex Shape Targets*, 16–22. Prairie View A&M University, Prairie View, Texas, Technical Report, NASA Grant Number: NGL44-033-017.

Uluisik, U., G. Cakir, M. Cakir, and L. Sevgi. 2008. "Radar Cross Section (RCS) Modeling and sSimulation, Part 1: A Tutorial Review of Definitions, Strategies, and Canonical Examples." *IEEE Antennas and Propagation Magazine* 50 (1): 115–126.

Chapter 5
Studies on Pre-selected Structures

The authenticity of the revised methodology is established in this chapter through several case studies. The results obtained from NEC have been compared with those available in open domain as well as with those obtained from full wave simulation software (FEKO). The segment length (Δ) and segment radius (r) in all the cases have been chosen in accordance with NEC modelling guidelines $\left(\Delta \gg \frac{\lambda}{10} \text{ and } \frac{\Delta}{r} > 8 \right)$.

5.1 Monostatic RCS of a PEC Dihedral

Canonical geometries are usually used for the validation of computational electromagnetics codes and PEC dihedral is one such characteristic target. The monostatic RCS of a 100 cm × 100 cm PEC dihedral (Sevgi 2003) at 600 MHz has been computed using NEC. The dihedral is assumed to be in the XY plane with the axis along Z-direction. The PEC dihedral has been modelled in HyperMesh using 2575 segments and the Nastran file generated has been converted to NEC format using an extraction algorithm. The wire grid model of PEC dihedral in HyperMesh is shown in Fig. 5.1 and the parameters *with respect to* simulation and meshing are summarized in Table 5.1. The NEC compatible segmentation data of PEC dihedral extracted from Nastran file and the NEC input file are shown in Figs. 5.2 and 5.3, respectively.

In the computation of monostatic RCS, for each value of azimuth angle (ϕ_{incident}), NEC generates the 360° radiation pattern. The RCS value corresponding to the specific ($\phi_{\text{scattering}}$) has been extracted from among all the available values. The results obtained from NEC have been validated *with respect to* FEKO and are found to be well matched, as shown in Fig. 5.4.

© The Author(s), under exclusive license to Springer Nature Singapore Pte Ltd. 2021
V. Joy et al., *Fundamentals of RCS Prediction Methodology using Parallelized Numerical Electromagnetics Code (NEC) and Finite Element Pre-processor*,
SpringerBriefs in Computational Electromagnetics,
https://doi.org/10.1007/978-981-15-7164-0_5

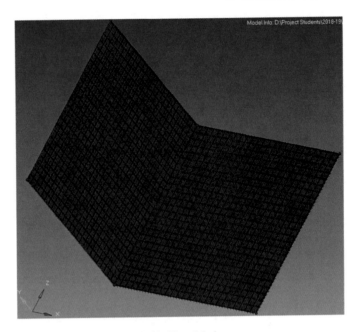

Fig. 5.1 Wire grid model of PEC dihedral in HyperMesh

Table 5.1 Simulation and meshing parameters to be used for the computation of monostatic RCS of PEC dihedral

Simulation parameters	
Frequency	600 MHz
Wavelength (λ)	0.5 m
Angles of illumination	$\theta_{incident} = 90°$; $\phi_{incident} = 0°-360°$
Polarization angle (η)	0°
Angles of scattering	$\theta_{scattering} = 90°$; $\phi_{scattering} = 0°-360°$
Meshing parameters	
Length of wire (m)	0.04
Radius of wire (m)	0.002
Clean-up tolerance	4e−4
Node tolerance	4e−5

5.2   Monostatic RCS of NASA Almond

NASA almond, a benchmark radar target with a doubly curved structure, is considered in this section. A metallic NASA almond of total length l (9.936 inches) can be modelled as per definitions in (Woo et al. 1993). The relevant equations describing the structure are given below:

NEC compatible segmentation data of PEC dihedral extracted from Nastran file							
GW 1	1	.6400	.0000	.2800	.6800	.0000	.2800	.0020
GW 1	1	.6400	.0000	.2800	.6400	.0000	.2400	.0020
GW 1	1	.6800	.0000	.2800	.6800	.0000	.2400	.0020
GW 1	1	.6800	.0000	.2400	.6400	.0000	.2400	.0020
.							
.							
.							
GW 1	1	.0000	.6400	.6000	.0000	.6400	.6400	.0020
GW 1	1	.0000	.6400	.6400	.0000	.6000	.6400	.0020
GW 1	1	.0000	.6400	.6400	.0000	.6400	.6800	.0020
GW 1	1	.0000	.6800	.6400	.0000	.6400	.6400	.0020

Fig. 5.2 A portion of the NEC compatible segmentation data, corresponding to PEC dihedral, extracted from Nastran file

NEC input file for the computation of monostatic RCS of PEC dihedral							
CM Monostatic RCS of PEC dihedral							
CM Frequency=600MHz; Wavelength=0.5m							
CM Angles of incidence: theta=90; phi=0-360; eta=0							
CM Angles of scattering: theta=90; phi=0-360							
CE							
GW 1	1	.6400	.0000	.2800	.6800	.0000	.2800	.0020
GW 1	1	.6400	.0000	.2800	.6400	.0000	.2400	.0020
.							
.							
.							
GW 1	1	.0000	.6400	.6400	.0000	.6400	.6800	.0020
GW 1	1	.0000	.6800	.6400	.0000	.6400	.6400	.0020
GE							
FR 0	1	0	0	600			
EX 1	1	721	0	90.	0	0	0.	0.5
RP 0	1	721	1000	90.	0	0.	0.5
EN							

Fig. 5.3 NEC input file for the computation of monostatic RCS of PEC dihedral at 600 MHz

- For $-0.416667 < t < 0$ and $-\pi < \mathrm{phi} < \pi$

$$x = (9.936)(t) \ \text{inches}$$

$$y = 0.193333 \times 9.936 \times \left(\sqrt{1 - \left(\frac{t}{0.416667}\right)^2}\right) \times \cos(\mathrm{phi})$$

$$z = 0.064444 \times 9.936 \times \left(\sqrt{1 - \left(\frac{t}{0.416667}\right)^2}\right) \times \sin(\mathrm{phi})$$

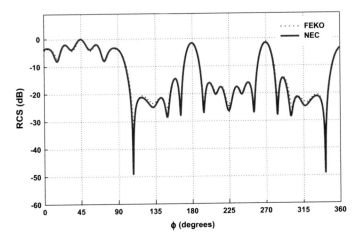

Fig. 5.4 Monostatic RCS of PEC dihedral

- For $0 < t < 0.58333$ and $-\pi < phi < \pi$

$$x = (9.936)(t) \ \ \text{inches}$$

$$y = 4.83345 \times 9.936 \times \left(\sqrt{1 - \left(\frac{t}{2.08335} \right)^2} - 0.96 \right) \times \cos(phi)$$

$$z = 1.61115 \times 9.936 \times \left(\sqrt{1 - \left(\frac{t}{2.08335} \right)^2} - 0.96 \right) \times \sin(phi)$$

The monostatic RCS of NASA almond at 1.19 GHz, modelled as per above definitions, has been computed using NEC. The metallic almond has been modelled and segmented in HyperMesh as shown in Fig. 5.5 using 3770 segments. The mesh data generated has been converted into NEC format and the required program control commands have been then inserted to frame the NEC input file. The problem definition *with respect to* simulation and meshing is summarized in Table 5.2. The NEC compatible segmentation data of metallic almond extracted from Nastran file and the NEC input file for horizontal and vertical polarizations are shown in Figs. 5.6, 5.7, and 5.8, respectively.

Figure 5.9 shows the monostatic RCS of metallic NASA almond obtained from NEC as well as FEKO at 1.19 GHz as a function of azimuth angle for both horizontal and vertical polarizations. The elevation angle has been fixed at 90° and azimuth angle has been varied from 0° to 180°. The results obtained from NEC as well as FEKO have been validated *w.r.t.* measured results in (Woo et al. 1993) and are found to be well matched.

Fig. 5.5 Wire grid model of NASA almond in HyperMesh

Table 5.2 Simulation and meshing parameters to be used for the computation of monostatic RCS of NASA almond

Simulation parameters	
Frequency	1.19 GHz
Wavelength (λ)	0.25 m
Angles of illumination	$\theta_{\text{incident}} = 90°$; $\phi_{\text{incident}} = 0° - 180°$
Polarization angle (η)	$0°$(for vertical polarization) $90°$(for horizontal polarization)
Angles of scattering	$\theta_{\text{scattering}} = 90°$; $\phi_{\text{scattering}} = 0° - 180°$
Meshing parameters	
Length of wire (m)	0.005
Radius of wire (m)	0.0005
Clean-up tolerance	5e−4
Node tolerance	5e−5

5.3 Monostatic RCS of Metallic Single Ogive

Another benchmark radar target used for the validation of electromagnetic analysis programs is a single ogive, which is a typical nosecone section in airborne platforms. The relevant equations required for modelling a single ogive are given below (Woo et al. 1993):

- For −5 in. $< t <$ 0 in. and $-\pi <$ phi $< \pi$,

If $g(t) = \left[\sqrt{\left(1 - \left(\frac{t}{5}\right)^2 \sin^2(22.62°)\right)} - \cos(22.62°) \right]$, then

$$y = \frac{g(t) \times \cos(\text{phi})}{1 - \cos(22.62°)}$$

NEC compatible segmentation data of NASA Almond extracted from Nastran file							
GW 1	1	-.1046	.0048	-.0007	-.1052	.0000	.0000	.0005
GW 1	1	-.1052	.0000	.0000	-.1046	.0049	-.0003	.0005
GW 1	1	-.1046	.0048	-.0007	-.1046	.0049	-.0003	.0005
GW 1	1	-.1029	.0094	-.0013	-.1046	.0048	-.0007	.0005
GW 1	1	-.1029	.0094	-.0013	-.1030	.0096	-.0006	.0005
.							
.							
.							
GW 1	1	.0365	-.0450	.0030	.0363	-.0424	.0059	.0005
GW 1	1	.0313	-.0457	.0030	.0312	-.0431	.0060	.0005
GW 1	1	.0261	-.0464	.0031	.0260	-.0437	.0060	.0005
GW 1	1	.0209	-.0469	.0031	.0208	-.0442	.0061	.0005
GW 1	1	.0157	-.0473	.0031	.0156	-.0446	.0062	.0005
GW 1	1	.0105	-.0476	.0032	.0104	-.0449	.0062	.0005
GW 1	1	.0052	-.0478	.0032	.0052	-.0450	.0062	.0005

Fig. 5.6 A portion of the NEC compatible segmentation data, corresponding to NASA almond, extracted from Nastran file

NEC input file for horizontal polarization
CM Monostatic RCS of metallic NASA almond
CM Frequency=1.19 GHZ; Wavelength= 0.25 m
CM Angles of incidence: theta=90 degree; phi=0 to 180 degree; eta=90
CM Angles of scattering: theta=90 degree; phi=0 to 180 degree
CE
GW 1 1 -.1046 .0048 -.0007 -.1052 .0000 .0000 .0005
GW 1 1 -.1052 .0000 .0000 -.1046 .0049 -.0003 .0005
.
.
GW 1 1 .0105 -.0476 .0032 .0104 -.0449 .0062 .0005
GW 1 1 .0052 -.0478 .0032 .0052 -.0450 .0062 .0005
GE
FR 0 1 0   0   1190
EX 1 1 181 0   90. 0. 90. 0. 1.
RP 0 1 181 1000 90. 0. 0. 1.
EN

Fig. 5.7 NEC input file for the computation of monostatic RCS of NASA almond (horizontal polarization)

$$z = \frac{g(t) \times \sin(\text{phi})}{1 - \cos(22.62°)}$$

- For 0 in. $< t < 5$ in. and $-\pi < \text{phi} < \pi$

If $f(t) = \left[\sqrt{\left(1 - \left(\frac{t}{5}\right)^2 \sin^2(22.62°)\right)} - \cos(22.62°) \right]$, then

```
NEC input file for vertical polarization

CM Monostatic RCS of metallic NASA almond
CM Frequency=1.19 GHz; Wavelength=0.25 m
CM Angles of incidence: theta=90 degree; phi=0 to 180 degree; eta=0
CM Angles of scattering: theta=90 degree; phi=0 to 180 degree
CE
GW  1   1   -.1046   .0048   -.0007   -.1052   .0000    .0000   .0005
GW  1   1   -.1052   .0000    .0000   -.1046   .0049   -.0003   .0005
 .

 .
GW  1   1    .0105  -.0476    .0032    .0104  -.0449    .0062   .0005
GW  1   1    .0052  -.0478    .0032    .0052  -.0450    .0062   .0005
GE
FR  0   1    0    0      1190
EX  1   1    181  0      90.    0.   0.   0.   1.
RP  0   1    181  1000   90.    0.   0.   1.
EN
```

Fig. 5.8 NEC input file for the computation of monostatic RCS of NASA almond (vertical polarization)

$$y = \frac{f(t) \times \cos(\text{phi})}{1 - \cos(22.62°)}$$

$$z = \frac{f(t) \times \sin(\text{phi})}{1 - \cos(22.62°)}$$

The metallic single ogive has been modelled and meshed in HyperMesh using 612 segments as shown in Fig. 5.10. The rest of the procedure is same as explained in the previous sections. The problem definition *with respect to* simulation and meshing is summarized in Table 5.3. The segmentation data of metallic single ogive extracted from Nastran file in accordance with NEC modelling guidelines is shown in Fig. 5.11. The NEC input files then framed for both horizontal and vertical polarizations are shown in Figs. 5.12 and 5.13, respectively.

Figure 5.14 shows the monostatic RCS of metallic single ogive at 1.18 GHz as a function of azimuth angle for both horizontal and vertical polarization. It is clear from the figure that the results obtained from NEC are well matched to the measured results in (Woo et al. 1993) as well with those obtained from FEKO.

5.4   Monostatic RCS of Rocket-Shaped Model

A hybrid structure composed of several canonical geometries is considered here (Bhushana et al. 2017). A rocket is one such structure and can be assumed to be made up of a cone, a cylinder, a frustum, and a circular disc as shown in Fig. 5.15. The metallic rocket-shaped model with dimensions as shown in Fig. 5.15 has been modelled and meshed in HyperMesh using 5678 segments. The wire grid model of

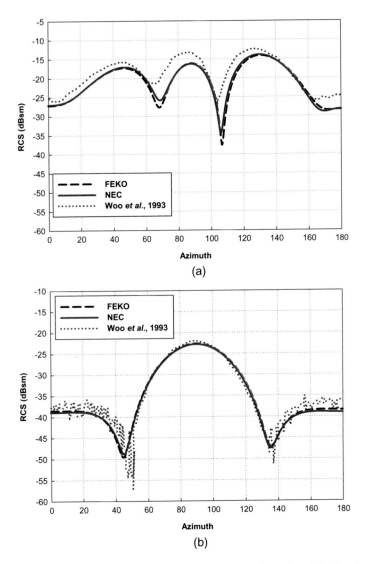

Fig. 5.9 Monostatic RCS of metallic NASA almond. **a** Horizontal polarization **b** vertical polarization

the same in HyperMesh is shown in Fig. 5.16 and the parameters *with respect to* simulation and meshing are summarized in Table 5.4. The rest of the procedure is same as explained in Chap. 4. The segmentation data of the rocket shaped model extracted from Nastran file in accordance with NEC modelling guidelines is shown in Fig. 5.17. The corresponding NEC input file is shown in Fig. 5.18.

Fig. 5.10 Wire grid model of metallic single ogive in HyperMesh

Table 5.3 Simulation and meshing parameters to be used for the computation of monostatic RCS of metallic single ogive

Simulation parameters	
Frequency	1.18 GHz
Wavelength (λ)	0.25 m
Angles of illumination	$\theta_{incident} = 90°$; $\phi_{incident} = 0° - 180°$
Polarization angle (η)	0° (for vertical polarization) 90° (for horizontal polarization)
Angles of scattering	$\theta_{scattering} = 90°$; $\phi_{scattering} = 0° - 180°$
Meshing parameters	
Length of wire (m)	0.01
Radius of wire (m)	0.0005
Clean-up tolerance	1e−4
Node tolerance	1e−5

The monostatic RCS of the hybrid model at 600 MHz as computed using both NEC as well as FEKO is shown in Fig. 5.19. It is apparent from the figure that the results from NEC are well in agreement with those from FEKO.

5.5 Monostatic RCS of Micro Aerial Vehicle (MAV)

The monostatic RCS of a flying wing MAV (Hassanalian and Abdelkefi 2017) at 600 MHz has been computed using NEC and the results have been validated with those from FEKO. The CAD model of metallic MAV in STEP format (.stp) as

NEC compatible segmentation data of metallic single ogive extracted from Nastran file							
GW 1	1	-.1178	.0041	.0000	-.1270	.0000	.0000	.0005
GW 1	1	-.1270	.0000	.0000	-.1178	.0035	-.0020	.0005
GW 1	1	-.1178	.0035	-.0020	-.1178	.0041	.0000	.0005
GW 1	1	-.1178	.0035	-.0020	-.1084	.0068	-.0039	.0005
GW 1	1	-.1084	.0068	-.0039	-.1084	.0078	.0000	.0005
.							
.							
.							
GW 1	1	.0302	.0132	.0229	.0302	.0229	.0132	.0005
GW 1	1	.0402	.0126	.0219	.0402	.0219	.0126	.0005
GW 1	1	.0502	.0119	.0206	.0502	.0205	.0119	.0005
GW 1	1	.0797	.0149	.0086	.0797	.0086	.0149	.0005
GW 1	1	.0601	.0109	.0190	.0601	.0189	.0109	.0005
GW 1	1	.0700	.0171	.0099	.0699	.0098	.0171	.0005

Fig. 5.11 A portion of the NEC compatible segmentation data, corresponding to metallic single ogive, extracted from Nastran file

NEC input file for horizontal polarization							
CM Monostatic RCS of metallic single ogive							
CM Frequency=1.18 GHz; Wavelength=0.25 m							
CM Angles of incidence: theta=90 degree; phi=0 to 180 degree; eta=90							
CM Angles of scattering: theta=90 degree; phi=0 to 180 degree							
CE							
GW 1	1	-.1178	.0041	.0000	-.1270	.0000	.0000	.0005
GW 1	1	-.1270	.0000	.0000	-.1178	.0035	-.0020	.0005
GW 1	1	-.1178	.0035	-.0020	-.1178	.0041	.0000	.0005
.							
.							
.							
GW 1	1	.0797	.0149	.0086	.0797	.0086	.0149	.0005
GW 1	1	.0601	.0109	.0190	.0601	.0189	.0109	.0005
GW 1	1	.0700	.0171	.0099	.0699	.0098	.0171	.0005
GE							
FR 0	1	0	0	1180			
EX 1	1	181	0	90. 0.	90. 0. 1.		
RP 0	1	181	1000	90. 0.	0. 1.		
EN							

Fig. 5.12 NEC input file for the computation of monostatic RCS of metallic single ogive (horizontal polarization)

shown in Fig. 5.20 has been imported from GrabCAD library (https://grabcad.com/library) into HyperMesh. GrabCAD is an online community where engineers and designers can view and share CAD files. The wire grid model of the metallic MAV in HyperMesh is shown in Fig. 5.21 and the parameters *with respect to* simulation and meshing are summarized in Table 5.5. The segmentation data for MAV, extracted from Nastran file, in accordance with NEC modelling guidelines is shown in Fig. 5.22.

NEC input file for vertical polarization

```
CM Monostatic RCS of metallic single ogive
CM Frequency=1.18 GHz; Wavelength=0.25 m
CM Angles of incidence: theta=90 degree; phi=0 to 180 degree; eta=0
CM Angles of scattering: theta=90 degree; phi=0 to 180 degree
CE
GW 1  1  -.1178   .0041   .0000  -.1270   .0000   .0000   .0005
GW 1  1  -.1270   .0000   .0000  -.1178   .0035  -.0020   .0005
GW 1  1  -.1178   .0035  -.0020  -.1178   .0041   .0000   .0005
GW 1  1  -.1178   .0035  -.0020  -.1084   .0068  -.0039   .0005
.
.
.
GW 1  1   .0601   .0109   .0190   .0601   .0189   .0109   .0005
GW 1  1   .0700   .0171   .0099   .0699   .0098   .0171   .0005
GE
FR 0  1   0   0   1180
EX 1  1  181   0    90. 0.   0. 0.  1.
RP 0  1  181  1000 90. 0.    0.  1.
EN
```

Fig. 5.13 NEC input file for the computation of monostatic RCS of metallic single ogive (vertical polarization)

The corresponding NEC input file is shown in Fig. 5.23.

Figure 5.24 shows the monostatic RCS of metallic MAV at 600 MHz, computed using both NEC as well as FEKO, as a function of azimuth angle. It is apparent that the NEC results are closely matching with those from FEKO.

5.6 Bi-Static RCS of Unmanned Aerial Vehicle (UAV)

The typical model of a UAV is considered in this section. The CAD model of metallic UAV as shown in Fig. 5.25 has been imported into HyperMesh in STEP format (*.stp) from GrabCAD. The meshed structure of metallic UAV in HyperMesh is shown in Fig. 5.26. The problem description *with respect to* simulation and meshing are summarized in Table 5.6. The segmentation data for the UAV extracted from Nastran file in accordance with NEC modelling guidelines is shown in Fig. 5.27. The corresponding NEC input file is shown in Fig. 5.28.

Figure 5.29 shows the bi-static RCS of the metallic UAV model at 150 MHz, computed using both NEC as well as FEKO, as a function of azimuth angle. The figure indicates good agreement between results obtained from NEC as well as FEKO.

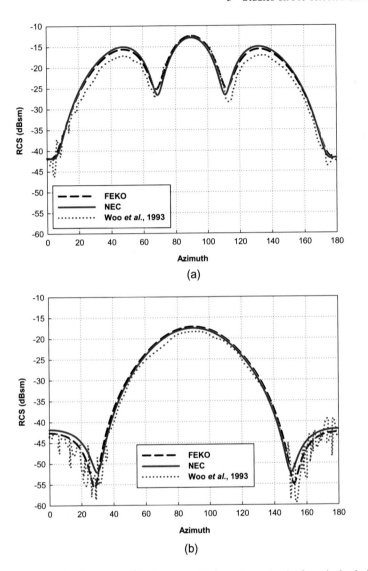

Fig. 5.14 Monostatic RCS of metallic single ogive. **a** Horizontal polarization **b** vertical polarization

5.7   Monostatic RCS of Automobile

The typical model of a car is considered in this section. The CAD model of the metallic car as shown in Fig. 5.30 has been imported into HyperMesh in IGES format (*.igs) from GrabCAD. The wire grid model of metallic car in HyperMesh is shown in Fig. 5.31. The problem description *with respect to* simulation and meshing are summarized in Table 5.7. The segmentation data for the car extracted from

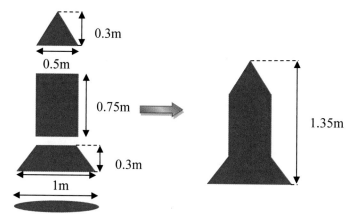

Fig. 5.15 Cross-sectional view of the metallic rocket-shaped model

Fig. 5.16 Wire grid model
of metallic rocket shaped
geometry in HyperMesh

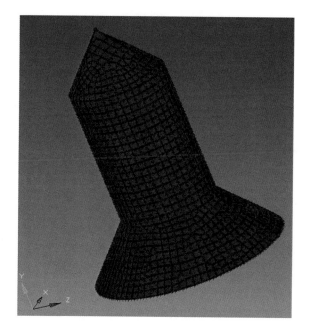

Nastran file in accordance with NEC modelling guidelines is shown in Fig. 5.32.
The corresponding NEC input file is shown in Fig. 5.33.

Figure 5.34 shows the monostatic RCS of the metallic car model at 1 GHz,
computed using both NEC as well as FEKO, as a function of azimuth angle. The
elevation angle has been fixed at 90° and azimuth angle has been varied from 0° to
360°. The figure indicates agreement between results obtained from NEC as well as
FEKO except for a slight difference in magnitude.

Table 5.4 Simulation and meshing parameters to be used for the computation of monostatic RCS of metallic rocket-shaped geometry

Simulation parameters	
Frequency	600 MHz
Wavelength (λ)	0.5 m
Angles of illumination	$\theta_{incident} = 90°$; $\phi_{incident} = 0°-360°$
Polarization angle (η)	0°
Angles of scattering	$\theta_{scattering} = 90°$; $\phi_{scattering} = 0° \theta_{scattering} = 90°$; $\phi_{scattering} = 0°-360°$
Meshing parameters	
Length of wire (m)	0.04
Radius of wire (m)	0.004
Clean-up tolerance	4e−4
Node tolerance	4e−5

NEC compatible segmentation data of metallic rocket shaped model extracted from Nastran file							
GW 1	1	-.2343	-.2152	-.3597	-.2187	-.1846	-.3395	.0040
GW 1	1	-.2343	-.2152	-.3597	-.2067	-.2205	-.3814	.0040
GW 1	1	-.2187	-.1846	-.3395	-.1954	-.1944	-.3627	.0040
GW 1	1	-.1954	-.1944	-.3627	-.2067	-.2205	-.3814	.0040
GW 1	1	-.2067	-.2205	-.3814	-.1761	-.2236	-.3992	.0040
GW 1	1	-.1954	-.1944	-.3627	-.1681	-.1982	-.3796	.0040
.							
.							
.							
GW 1	1	.1560	.8590	-.0318	.1339	.8876	-.0201	.0040
GW 1	1	.1135	.9001	-.0523	.1339	.8876	-.0201	.0040
GW 1	1	.1339	.8876	-.0201	.1173	.9091	.0062	.0040
GW 1	1	.1480	.8722	.0079	.1339	.8876	-.0201	.0040
GW 1	1	.1173	.9091	.0062	.1328	.8872	.0276	.0040
GW 1	1	.1187	.8966	.0475	.1328	.8872	.0276	.0040

Fig. 5.17 A portion of the NEC compatible segmentation data, corresponding to metallic rocket shaped model, extracted from Nastran file

NEC input file for the computation of monostatic RCS of rocket shaped model

```
CM Monostatic RCS of Rocket shaped model
CM Frequency=600 MHz; Wavelength=0.5 m
CM Angles of incidence: theta=90 degree; phi=0 to 360 degree; eta=0
CM Angles of scattering: theta=90 degree; phi=0 to 360 degree
CE
GW  1   1   -.2343   -.2152   -.3597   -.2187   -.1846   -.3395    .0040
GW  1   1   -.2343   -.2152   -.3597   -.2067   -.2205   -.3814    .0040
GW  1   1   -.2187   -.1846   -.3395   -.1954   -.1944   -.3627    .0040
  .
  .
  .
GW  1   1    .1480    .8722    .0079    .1339    .8876   -.0201    .0040
GW  1   1    .1173    .9091    .0062    .1328    .8872    .0276    .0040
GW  1   1    .1187    .8966    .0475    .1328    .8872    .0276    .0040
GE
FR  0   1   0   0   600
EX  1   1   361 0   90.   0.   0.   0.   1.
RP  0   1   361 1000 90.   0.   0.   1.
EN
```

Fig. 5.18 NEC input file for the computation of monostatic RCS of rocket-shaped model

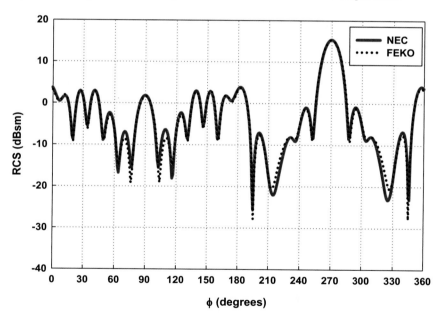

Fig. 5.19 Monostatic RCS of metallic rocket shaped model

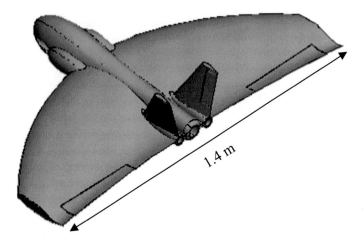

Fig. 5.20 CAD model of metallic MAV

Fig. 5.21 Wire grid model of Metallic MAV in HyperMesh

Table 5.5 Simulation and meshing parameters to be used for the computation of monostatic RCS of metallic MAV

Simulation parameters	
Frequency	600 MHz
Wavelength (λ)	0.5 m
Angles of illumination	$\theta_{incident} = 90°$; $\phi_{incident} = 0°-360°$
Polarization angle (η)	0°
Angles of scattering	$\theta_{scattering} = 90°$; $\phi_{scattering} = 0°-360°$
Meshing parameters	
Length of wire (m)	0.04
Radius of wire (m)	0.004
Clean-up tolerance	4e−4
Node tolerance	4e−5

NEC compatible segmentation data of metallic MAV extracted from Nastran file							
GW 1	1	-.0700	.3738	.0200	-.0691	.4196	.0435	.0040
GW 1	1	-.0700	.3738	.0200	-.0705	.3725	.0200	.0040
GW 1	1	-.0691	.4196	.0435	-.0795	.3934	.0559	.0040
GW 1	1	-.0795	.3934	.0559	-.0705	.3725	.0200	.0040
GW 1	1	-.0700	.3738	.0200	-.0624	.4078	.0185	.0040
.							
.							
.							
GW 1	1	-.0706	.3751	.0197	-.0700	.4085	.0200	.0040
GW 1	1	-.0700	.4085	.0200	-.0856	.4072	.0125	.0040
GW 1	1	-.0849	.4430	.0133	-.0700	.4431	.0200	.0040
GW 1	1	-.0823	.4787	.0157	-.0700	.4778	.0200	.0040
GW 1	1	-.0791	.5144	.0178	-.0700	.5125	.0200	.0040

Fig. 5.22 A portion of the NEC compatible segmentation data, corresponding to metallic MAV, extracted from Nastran file

NEC input file for the computation of monostatic RCS of metallic MAV

```
CM Monostatic RCS of metallic MAV
CM Frequency=600 MHz; Wavelength=0.5 m
CM Angles of incidence: theta=90 degree; phi=0 to 360 degree; eta=0
CM Angles of scattering: theta=90 degree; phi=0 to 360 degree
CE
GW 1  1  -.0700   .3738   .0200   -.0691   .4196   .0435   .0040
GW 1  1  -.0700   .3738   .0200   -.0705   .3725   .0200   .0040
GW 1  1  -.0691   .4196   .0435   -.0795   .3934   .0559   .0040
.

.
GW 1  1  -.0823   .4787   .0157   -.0700   .4778   .0200   .0040
GW 1  1  -.0791   .5144   .0178   -.0700   .5125   .0200   .0040
GE
FR 0  1   0   0   600
EX 1  1   361  0   90. 0. 0.  0.  1.
RP 0  1   361  1000 90. 0.  0.   1.
EN
```

Fig. 5.23 NEC input file for the computation of monostatic RCS of metallic MAV

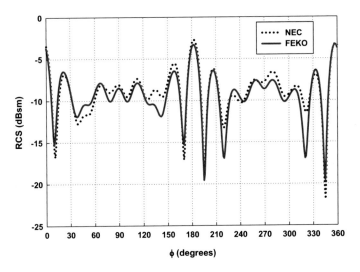

Fig. 5.24 Monostatic RCS of metallic MAV

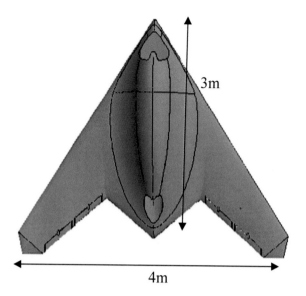

Fig. 5.25 CAD model of metallic UAV

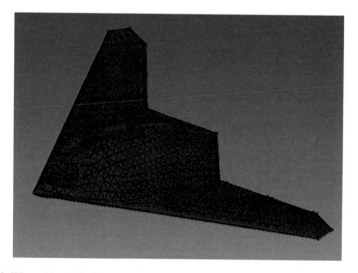

Fig. 5.26 Wire grid model of metallic UAV in HyperMesh

Table 5.6 Simulation and meshing parameters to be used for the computation of bi-static RCS of metallic UAV

Simulation parameters	
Frequency	150 MHz
Wavelength (λ)	2 m
Angles of illumination	$\theta_{incident} = 90°$; $\phi_{incident} = 90°$
Polarization angle (η)	90°
Angles of scattering	$\theta_{scattering} = 0°-360°$; $\phi_{scattering} = 90°$
Meshing parameters	
Length of wire (m)	0.1
Radius of wire (m)	0.01
Clean-up tolerance	1e−6
Node tolerance	1e−7

NEC compatible segmentation data for metallic UAV extracted from Nastran file							
GW 1	1	-.8631	.0344	-.3568	-.9184	.0357	-.4394	.0100
GW 1	1	-.8166	.0297	-.4485	-.9184	.0357	-.4394	.0100
GW 1	1	-.8166	.0297	-.4485	-.8705	.0312	-.5290	.0100
GW 1	1	-.9184	.0357	-.4394	-.8705	.0312	-.5290	.0100
GW 1	1	-.8705	.0312	-.5290	-.9728	.0370	-.5219	.0100
GW 1	1	-.9184	.0357	-.4394	-.9728	.0370	-.5219	.0100
.							
.							
.							
GW 1	1	-.5209	.0076	-.5369	-.6041	.0151	-.5875	.0100
GW 1	1	-.5128	.0113	-.7354	-.6020	.0171	-.6849	.0100
GW 1	1	-.5148	.0089	-.6365	-.6020	.0171	-.6849	.0100
GW 1	1	-.5106	.0144	-.8336	-.5977	.0196	-.7835	.0100
GW 1	1	-.5128	.0113	-.7354	-.5977	.0196	-.7835	.0100
GW 1	1	-.5937	.0230	-.8799	-.5977	.0196	-.7835	.0100
GW 1	1	-.5937	.0230	-.8799	-.6844	.0281	-.8278	.0100
GW 1	1	-.8631	.0344	-.3568	-.8166	.0297	-.4485	.0100

Fig. 5.27 A portion of the NEC compatible segmentation data, corresponding to metallic UAV, extracted from Nastran file

NEC input file for the computation of bi-static RCS of metallic UAV							
CM Bi-static RCS of UAV							
CM Frequency=150 MHz; Wavelength=2 m							
CM Angles of incidence: theta=90 degree; phi=90 degree; eta=90							
CM Angles of scattering: phi=90 degree; theta=0 to 360 degree							
CE							
GW 1	1	-.8631	.0344	-.3568	-.9184	.0357	-.4394	.0100
GW 1	1	-.8166	.0297	-.4485	-.9184	.0357	-.4394	.0100
GW 1	1	-.8166	.0297	-.4485	-.8705	.0312	-.5290	.0100
.							
.							
GW 1	1	-.5937	.0230	-.8799	-.6844	.0281	-.8278	.0100
GW 1	1	-.8631	.0344	-.3568	-.8166	.0297	-.4485	.0100
GE							
FR 0	1	0	0	150			
EX 1	1	1	0	90.	90.	90.	
RP 0	361	1	1000 0.	90.	1.	0.	
EN							

Fig. 5.28 NEC input file for the computation of bi-static RCS of metallic UAV

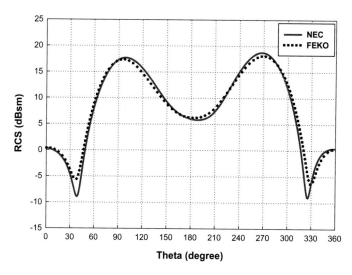

Fig. 5.29 Bi-static RCS of metallic UAV model

Fig. 5.30 CAD model of metallic car

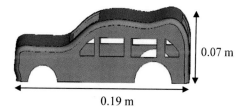

0.07 m

0.19 m

Fig. 5.31 Wire grid model of metallic car in HyperMesh

Table 5.7 Simulation and meshing parameters to be used for the computation of monostatic RCS of metallic car

Simulation parameters	
Frequency	1 GHz
Wavelength (λ)	0.3 m
Angles of illumination	$\theta_{incident} = 90°$; $\phi_{incident} = 0°-360°$
Polarization angle (η)	0°
Angles of scattering	$\theta_{scattering} = 90°$; $\phi_{scattering} = 0°-360°$
Meshing parameters	
Length of wire (m)	0.005
Radius of wire (m)	0.0005
Clean-up tolerance	5e−6
Node tolerance	5e−7

NEC compatible segmentation data of metallic car extracted from Nastran file								
GW 1	1	.2693	-.0129	.0071	.2693	-.0168	.0069	.0005
GW 1	1	.2693	-.0129	.0071	.2673	-.0129	.0071	.0005
GW 1	1	.2693	-.0168	.0069	.2673	-.0168	.0069	.0005
GW 1	1	.2673	-.0168	.0069	.2673	-.0129	.0071	.0005
GW 1	1	.2693	-.0168	.0069	.2693	-.0207	.0066	.0005
GW 1	1	.2693	-.0207	.0066	.2673	-.0207	.0066	.0005
GW 1	1	.2673	-.0207	.0066	.2673	-.0168	.0069	.0005
GW 1	1	.0745	-.0652	.0015	.0748	-.0649	-.0034	.0005.
.								
.								
.								
GW 1	1	.1915	-.0692	.0214	.1909	-.0703	.0224	.0005
GW 1	1	.1965	-.0692	.0214	.1955	-.0703	.0224	.0005
GW 1	1	.1965	-.0695	.0266	.1955	-.0705	.0270	.0005
GW 1	1	.1965	-.0698	.0317	.1955	-.0707	.0317	.0005
GW 1	1	.1965	-.0700	.0369	.1955	-.0710	.0363	.0005

Fig. 5.32 A portion of the NEC compatible segmentation data, corresponding to metallic car, extracted from Nastran file

NEC input file for the computation of monostatic RCS of metallic car								
CM Monostatic RCS of metallic car								
CM Frequency=1 GHz; Wavelength=0.3 m								
CM Angles of incidence: theta=90 degree; phi=0 to 360 degree; eta=0								
CM Angles of scattering: theta=90 degree; phi=0 to 360 degree								
CE								
GW 1	1	.2693	-.0129	.0071	.2693	-.0168	.0069	.0005
GW 1	1	.2693	-.0129	.0071	.2673	-.0129	.0071	.0005
GW 1	1	.2693	-.0168	.0069	.2673	-.0168	.0069	.0005
.								
.								
GW 1	1	.1965	-.0698	.0317	.1955	-.0707	.0317	.0005
GW 1	1	.1965	-.0700	.0369	.1955	-.0710	.0363	.0005
GE								
FR 0	1	0	0	1000				
EX 1	1	361	0	90	0. 0.	0.	1.	
RP 0	1	361	1000	90.	0. 0.	1.		
EN								

Fig. 5.33 NEC input file for the computation of monostatic RCS of metallic car

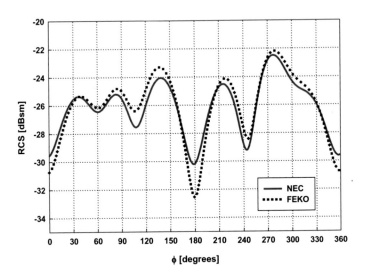

Fig. 5.34 Monostatic RCS of metallic car model

References

Bhushana, G.S., S. Nambari, S. Kota, and K.S. Ranga Rao. 2017. "Monostatic Radar Cross
    SECTION Estimation of Missile Shaped Object Using Physical Optics Method." In *IOP Confer-
    ence Series: Materials Science and Engineering* 7. https://doi.org/10.1088/1757-899x/225/1/
    012278.
Hassanalian, M., and A. Abdelkefi. 2017. "Classifications, Applications, and Design Challenges of
    Drones: A review." *Article in Progress in Aerospace Sciences*. 33. https://doi.org/10.1016/j.pae
    rosci.2017.04.003.
Sevgi, L. 2003. *Complex Electromagnetic Problems and Numeric Simulation Approaches*. Wiley-
    IEEE Press, 179–218. ISBN: 978-0-471-43062-9.
Woo, A.C., H.T.G. Wang, M.J. Schuh, and M.L. Sanders. 1993. EM Programmer's Notebook,
    "Benchmark Radar Targets for the Validation Computational Electromagnetics Programs." *IEEE
    Antennas and Propagation Magazine* 35(1): 84–89.

Chapter 6
Conclusion

The computation of RCS of airborne structures plays a crucial role in the development of stealth platforms, a pre-requisite currently in strategic sector. Accurate prediction of RCS incorporating all associated electromagnetic phenomena still remains a herculean task at high frequencies. In this regard, the present book has focused on scattering predictions using a parallelized version of method of moment-based Numerical Electromagnetics Code (NEC), a highly precise and resourceful tool for electromagnetic analysis. The theoretical framework of the open-source program starting from the formulation of integral equations to the solution of matrix equation has been briefly explained. The strategy for parallelization including the details on the application of ScaLAPACK and MPI has also been covered. A novel feature introduced in this book is the application of Altair HyperMesh, a multi-disciplinary finite element pre-processor, for the generation of NEC input files. This has greatly simplified one of the most intricate and error prone step in NEC, i.e. the translation of geometry data into a format compatible with NEC guidelines. The revised methodology has been then used to predict the RCS of various benchmark radar targets (dihedral, NASA almond, ogive, etc.) as well as complex CAD models. The results obtained show good agreement with those in literature as well with those simulated using commercial software, thereby establishing the authenticity of the revised methodology.

V. Joy et al., *Fundamentals of RCS Prediction Methodology using Parallelized Numerical Electromagnetics Code (NEC) and Finite Element Pre-processor*, SpringerBriefs in Computational Electromagnetics, https://doi.org/10.1007/978-981-15-7164-0_6

Appendix A
Libraries used for Parallel Computations

The effectiveness of the parallelized version of NEC largely depends on the choice of the operating system and the appropriate scientific libraries. A collection of popular libraries used for high-performance computing can be found at https://www.netlib.org. These specialized libraries are usually configured for the particular operating system and hardware platform on which the code is to be executed. Nowadays, most of the computer manufacturers build their own libraries so as to match the software with the hardware.

The CSIR 4PI high-performance computing facility (360 TF super computer) along with Intel cluster studio has been chosen as the computing platform for the work presented in this book. Intel cluster studio is a multi-component software toolkit with all the core libraries and tools required to efficiently develop and distribute parallel programs in clusters with Intel processors. ScaLAPACK, a parallel library package based on message-passing with MPI, has also been used in addition to this for implementing two-dimensional block-cyclic decomposition. Furthermore, Open MPI has been used to efficiently compile and launch the parallel jobs on clusters with the Linux operating systems.

A brief overview of ScaLAPACK and message-passing interface standard (MPI) is presented in the following sections.

A.1 ScaLAPACK

The two-dimensional block-cyclic decomposition used in (Rubinstein et al. 2003) hasbeen implemented using ScaLAPACK (Scalable Linear Algebra Package, or Scalable LAPACK) library. ScaLAPACK is a library of high-performance linear algebra routines for distributed memory message-passing computers and networks of heterogeneous machines (Blackford et al. 1997; Kontoghiorghes et al. 2005). ScaLAPACK can solve least squares problems, dense and banded linear systems, singular value

V. Joy et al., *Fundamentals of RCS Prediction Methodology using Parallelized Numerical Electromagnetics Code (NEC) and Finite Element Pre-processor*, SpringerBriefs in Computational Electromagnetics, https://doi.org/10.1007/978-981-15-7164-0

problems, eigenvalue problems, matrix factorizations, and estimation of the condition number of a matrix. ScaLAPACK uses block-partitioned algorithms to achieve high levels of data reuse. Furthermore, it employs well-crafted low-level modular components for simplifying the parallelization of high level routines by making their source code similar to that of the sequential one. The library is predominantly written in Fortran except for a few symmetric eigen problem auxiliary routines written in C. It is basically a software package provided by University of Tennessee, University of California, Berkeley, University of Colorado Denver, and NAG Ltd.

The concept of library-based parallelization began with the development of basic linear algebra subroutines (BLAS) which is basically a collection of Fortran 77 subroutines capable of performing basic linear algebra operations. At present, BLAS is classified into three sets of routines called "level" as given below:

- Level 1 BLAS-For performing scalar, vector, and vector–vector operations
- Level 2 BLAS-For performing matrix-vector operations
- Level 3 BLAS-For performing matrix–matrix operations.

The BLAS routines were then used for the development of the LAPACK library which predominantly used level 3 BLAS kernels. LAPACK was mainly designed to work with single-processor high-performance computers with vector processors and shared memory parallel computers. The requirement to solve similar problems on distributed memory architectures led to the development of basic linear algebra communication subroutines (BLACS). BLACS consists of a set of machine-independent and portable communication subroutines for performing linear algebra operations on distributed memory platforms. During the execution of BLACS, the processes involved are arranged in two-dimensional grids and each process can be referenced by its coordinates on the grid. Furthermore, each process is treated as a processor and the execution of one process can affect other processes only through the use of message passing.

Further, parallel basic linear algebra communication subprograms (PBLAS) have been developed. PBLAS is basically a collection of distributed linear algebra operations (similar to the sequential BLAS). The main aim of PBLAS is to distribute matrices among BLACS processes and then use BLACS as the communication platform. Subsequently, ScaLAPACK has been introduced for performing linear algebra operations on distributed memory message-passing computers and heterogeneous clusters. The functionality of ScaLAPACK for distributed memory architectures is analogous to that of LAPACK for workstations, vector supercomputers, and shared memory parallel computers. ScaLAPACK is based on BLACS and PBLAS packages, whereas LAPACK is based on BLAS routines.

The major objectives of the ScaLAPACK project are efficiency (to accelerate the program), portability (across all significant parallel machines), flexibility (to enable users to develop new routines from well-designed ones), scalability (as the number of processors and problem size grow), reliability (including error bounds), and ease of use (by having similar interfaces to LAPACK and ScaLAPACK). The machine dependencies are restricted to BLAS, LAPACK, and BLACS. LAPACK can be used on any machine where BLAS is available, and ScaLAPACK on any machine where BLAS, LAPACK, and BLACS are available (Kontoghiorghes et al. 2005).

A.2 Message-Passing Interface Standard (MPI)

Message Passing Interface Standard (MPI) is a library specification for message passing based on the suggestions of the MPI Forum which has over 40 participating establishments including industries, academicians, and users (Barney et al. 2018). The primary aim of this standard is to layout a set of efficient, portable, and flexible guidelines for message passing that can be extensively used for writing message-passing programs. MPI is not a library by itself but a description of what such a library should be. Although MPI is not an IEEE or ISO standard, it has become a widely accepted vendor independent industry standard for writing message-passing programs on high-performance computing platforms including distributed memory systems, shared memory systems, etc. However, the developer is responsible for correctly identifying the parallelization strategy and implementing the same using MPI constructs. Out of the several MPI libraries available, Open MPI has been used in this book. A general MPI program structure is shown in Figure A.1.

A.3 Procedure for Compilation and Execution of the Parallelized Version of NEC on Linux Platforms

The complete source code for the parallelized version of NEC can be found on https://github.com/vineethajoy/Parallel-NEC. The program has been compiled and executed on the CSIR 4PI high-performance computing facility (360 TF super computer) according to the following procedure:

Step 1: Install gcc4.8.5 (GNU Compiler Collection) which has gfortran (Fortran 95/2003/2008 compiler).

Step 2: Install Intel cluster studio 2013.

Step 3: Install ScaLAPACK, BLAS, and LAPACK library packages.

Step 4: Install Open MPI
         (version: openmpi-3.0; source: https://www.open-mpi.org)

Step 5: Copy the code (nec.f) and place it inside a folder. The NEC input file has to be copied into the same folder. The code has to be then compiled using the following command in the command window:

         gfortran −Og −g nec.f −o necout −lscalapack −lblas
         −lmpi

where necout is the executable file.

Fig. A.1 Structure of a
general MPI program

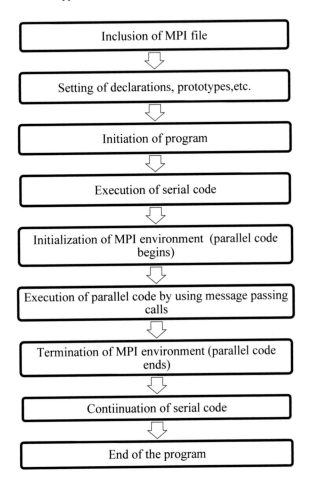

Step 6: Use the mpirun command as given below to launch the code.

mpirun −np < no: of processes> < necout>

Open MPI extracts information regarding the scheduling of processes on nodes
from the *mpirun* command line (Gabriel et al. 2004). The number of processes to
be launched is directly given by the -*np* switch. The strategy regarding where the
processes are to be launched depends on the list of nodes, the scheduling policy,
and the maximum number of slots available on each node. Slots denote the number
of processors available on a particular node. A default value of 1 is assumed if the
number of slots on a particular system is not explicitly mentioned. However, a precise
slot count is automatically provided by schedulers like PBS. Open MPI schedules
processes either by slot or by node. In the former mode, processes will be scheduled
on a particular node until all of its default slots are utilized. The next node will be

considered once this is completed. In the latter mode, all the nodes will be assigned a single process based on round-robin scheduling until all processes are scheduled. When all the default slots of a particular node are utilized, it will be skipped.

The CSIR 4PI high-performance computing facility uses PBSPro workload management software for job scheduling. Therefore, the jobs have been submitted to the HPC using a Portable Batch System (PBS) script containing information on the computational resources requested for the calculation, as well as the required commands for executing the program. A sample PBS script is shown in Fig. A.2.

In the case of nodes with processors containing multiple cores, Open MPI schedules consecutive processes on adjacent processor cores. This implies that a separate core will be allotted to each process. For instance, consider a node with 4 processors each containing 4 cores. In such a case, 16 processes can be run simultaneously at full speed, with one process running on a single core. On the other hand, a node with 4 single core processors can handle only 4 processes at maximum efficiency. Therefore, the number of processes to be launched needs to be judiciously determined based on the problem size under consideration and the number of cores available for utilization.

PBS script

```
#!/bin/csh
#PBS -l walltime=00:30:00
#PBS -l select=1: ncpus=4: mpiprocs=4
# Go to the directory from which you submitted the job
cd / NEC/ CODE/
# Load the required libraries
module load scalapack-2.0.2_shared
module load mpc-1.0.1
module load mpfr-3.1.4
module load gcc4.8.5
module load zlib-1.2.11
module load openmpi-3.0
module load blas-3.8.0
module load intel-cluster-studio-2013
gfortran -Og -g nec.f -o necout -lscalapack -lblas -lmpi
mpirun -np 16 / NEC/ CODE/necout
```

Fig. A.2 Format of a typical PBS script

Appendix B
Format of Command Lines in NEC

B.1 GW Command

Purpose: To generate a wire with an arbitrary number of segments.

Format:

Identifier	Int-1	Int-2	Re-1	Re-2	Re-3	Re-4	Re-5	Re-6	Re-7
GW	ITAG	N	X1	Y1	Z1	X2	Y2	Z2	RADIUS
	Tag number assigned to all the constituent wire segments	Number of segments into which the wire is to be divided	Co-ordinates of wire end 1			Co-ordinates of wire end 2			Wire radius

Example: GW 1 50 −0.05 0 −0.05 0.05 0 −0.05 0.02

Description: The above line generates a wire with a radius of 0.02 between (−0.05, 0, −0.05) and (0.05, 0, −0.05).

The wire will also be divided into 50 segments with each segment having tag number 1.

Notes:

- A segment is identified by its tag number and the number of the segment in the set of segments having that particular tag. The tag number is used for identifying a segment in the model when it is to be referenced later. If no tag number is to be given, the field can be left blank. Any number except zero can be used as a tag.

© The Author(s), under exclusive license to Springer Nature Singapore Pte Ltd. 2021
V. Joy et al., *Fundamentals of RCS Prediction Methodology using Parallelized Numerical Electromagnetics Code (NEC) and Finite Element Pre-processor*,
SpringerBriefs in Computational Electromagnetics,
https://doi.org/10.1007/978-981-15-7164-0

- The direction from end 1 to end 2 on each segment is taken as the positive reference direction for current.
- The input quantities are to be given in meters and if it is not, then they must be scaled to meters through a GS (Scale Structure Dimensions) command.

B.2   FR Command

Purpose: To specify the frequency of operation in Megahertz (MHz)

Format:

Identifier	Int-1	Int-2	Int-3	Int-4	Re-1	Re-2	Re-3	Re-4	Re-5	Re-6
FR	IFR	NFR	–	–	FR	DELFR	–	–	–	–
	Type of frequency stepping	No: of frequency steps	Blank columns		Frequency in MHz	Frequency stepping increment	Blank columns			

Example: FR 0 1 0 0 600

Description: The above line sets the operating frequency as 600 MHz.

Notes:

- The default frequency in NEC is 299.8 MHz which will be automatically set if a frequency command does not appear in the input file.
- In case frequency commands are grouped together, only the information on the last command line will be used.
- If an FR command with NFR greater than 1 is read, only an RP or XQ command will initiate the execution.

B.3   EX Command

Purpose: This command is used to mention the excitation for the structure. It can be an incident plane-wave of linear or elliptical polarization, a voltage source or an elementary current source.

Format:

Identifier	Int-1	Int-2	Int-3	Int-4	Re-1	Re-2	Re-3	Re-4	Re-5	Re-6
EX										

Table B.1 Options available
for Int-1 (EX command)

Int-1	Interpretation
0	Denotes a voltage source (Applied E-field source)
1	Denotes a linearly polarized incident plane wave
2	Denotes a right-hand elliptically polarized incident plane wave
3	Denotes an elementary current source
4	Denotes an elementary current source
5	Denotes a voltage source (current-slope-discontinuity)

The various parameters are described below:

1. Int-1: This integer value defines the type of excitation and it can take different values as shown in Table B.1.
   The remaining integers depend on the type of excitation chosen. Since this report deals with only plane wave excitation i.e. Int-1 = 1, 2 or 3, only those details are given here.
2. Int-2: Number of elevation (θ) angles required for the incident plane wave.
3. Int-3: Number of azimuth (ϕ) angles required for the incident plane wave.
4. Int-4: A value of 1 for this integer will initiate the computation and display of maximum relative admittance matrix asymmetry for network connections. 0 for Int-1 indicates that no action is to be taken.
5. Re-1: θ (of incident plane wave) in degrees.
6. Re-2: ϕ (of incident plane wave) in degrees.
7. Re-3: Polarization angle of incident plane wave (η) in degrees. In case of linear polarization, η denotes the angle between the unit vector along theta direction and the direction of the electric field. For elliptical polarization, η represents the angle between the unit vector along theta direction and the major ellipse axis.
8. Re-4: Step size for elevation angle (θ) increments in degrees.
9. Re-5: Step size for azimuth angle (ϕ) increments in degrees.
10. Re-6: Ratio of minor axis to major axis for elliptic polarization.

B.4 RP Command

Purpose: To mention sampling parameters for the computation of radiation pattern and initiate program execution.

Format:

Identifier	Int-1	Int-2	Int-3	Int-4	Re-1	Re-2	Re-3	Re-4	Re-5	Re-6
RP										

Table B.2 Options available for Int-1 (RP command)

Int-1	Interpretation
0	Space-wave fields will be calculated. If an infinite ground plane has been specified by a GN command, it will be included else antenna will be considered to be in free space
1	The surface waves travelling along the ground (ground parameters should be mentioned on a GN command line) will be added to the normal space wave
2	Linear cliff with antenna above upper level
3	Circular cliff with center at the origin of coordinate system and antenna above upper level
4	Radial wire ground screen with center at origin
5	Both linear cliff and radial wire ground screen
6	Both circular cliff and radial wire ground screen

The various parameters are described below:

1. Int-1: This integer chooses the way in which the radiated field will be calculated. Some values of Int-1 will affect the meaning of the remaining parameters on the command line. Options available for Int-1 are summarized in Table B.2. Options 2–6 initiate the computation of space waves with special ground conditions which have to be initialized on a GN or GD command line before the RP command is read.
2. Int-2: Number of values of elevation angle (θ) at which the field is to be evaluated.
3. Int-3: Number of values of azimuth angle (ϕ) at which field is to be evaluated.
4. Int-4 (PNGA): This integer should be interpreted as four independent numbers each having a different meaning. Int-4 will be irrelevant if Int-1 is equal to 1 and should be left blank in such case.

   - P: This integer determines the format in which output will be displayed. The options available are summarized in Table B.3.
   - N—In addition to the normal gain output, this integer controls the display of normalized gain for user defined field points. The normalization can be w.r.t. either its maximum value or a value specified by the user on field Re-6. The options available for this integer are summarized in Table B.4.
   - G—The value of this integer determines whether power gain or directive gain is to be chosen for display as well as normalization. In the case of plane wave excitation, radar cross-section will be printed under the section "gain" and therefore will not be affected by this value. The options available for this integer are summarized in Table B.5.

Table B.3 Options available for P (RP command)

P	Interpretation
0	Major axis, minor axis and total gain will be displayed
1	Vertical, horizontal and total gain will be displayed

Table B.4 Options available for N (RP command)

N	Interpretation
0	No normalized gain
1	Gain along major axis will be normalized
2	Gain along minor axis will be normalized
3	Gain along vertical axis will be normalized
4	Gain along horizontal axis will be normalized
5	Total gain will be normalized

Table B.5 Options available for G (RP command)

G	Interpretation
0	Power gain will be chosen
1	Directive gain will be chosen

Table B.6 Options available for A (RP command)

A	Interpretation
0	No averaging
1	Average gain will be evaluated
2	Average gain will be evaluated and the gain at field points used for averaging will not be displayed. Average gain will not be evaluated for any value of A If NTH or NPH is equal to one as the area of the region covered by field points vanishes

- A—This integer initiates the computation of average power gain over the region where fields have been evaluated. The options available are summarized in Table B.6.

5. Re-1: Initial elevation (θ) angle in degrees
6. Re-2: Initial azimuth (ϕ) angle in degrees.
7. Re-3: Step size for elevation (θ) angle in degrees
8. Re-4: Step size for azimuth (ϕ) angle in degrees
9. Re-5: Radial distance (R) of the field point from the origin (in meters).
10. Re-6: This field determines the value to which the gain will be normalized as per request on Int-4.

Example: RP 0 1 181 1000 90.0 0 0 1

Description: Plots the radiation pattern in the $\theta = 90°$ plane with ϕ varying from $0°$ to $180°$ in steps of $1°$.

Appendix C
Format of Various Output Files

C.1 Output from HyperMesh in Nastran Format for NASA Almond

```
$$------------------------------------------------------------------------------------$
$$                                                                                    $
$$ NASTRAN Input Deck Generated by HyperMesh Version  :2017.0.0.24
$$ Generated using HyperMesh-Nastran Template Version: 2017
$$                                                                                    $
$$   Template: NastranMSC general                                                     $
$$                                                                                    $
$$------------------------------------------------------------------------------------$
$$------------------------------------------------------------------------------------$
$$              Executive Control Cards                                               $
$$------------------------------------------------------------------------------------$
CEND
$$------------------------------------------------------------------------------------$
$$              Case Control Cards                                                    $
$$------------------------------------------------------------------------------------$
$$------------------------------------------------------------------------------------$
$$              Bulk Data Cards                                                       $
$$------------------------------------------------------------------------------------$
BEGIN BULK
$$
$$  GRID Data
$$
GRID         1            -.1045524.8252-3-6.662-4
GRID         2            -.1051533.0207-5-2.003-6
GRID         3            -.1045974.9269-3-3.267-4
GRID         4            -0.102859.3889-3-1.296-3
GRID         5            -.1030079.6266-3-6.383-4
```

© The Author(s), under exclusive license to Springer Nature Singapore Pte Ltd. 2021
V. Joy et al., *Fundamentals of RCS Prediction Methodology using Parallelized Numerical Electromagnetics Code (NEC) and Finite Element Pre-processor*,
SpringerBriefs in Computational Electromagnetics,
https://doi.org/10.1007/978-981-15-7164-0

```
GRID      6         -.1002951.3548-2-1.871-3
GRID      7         -.100582 1.396-2-9.256-4
GRID      8         -9.713-21.7274-2-2.385-3
GRID      9         -9.754-2.0178858-1.186-3
GRID     10         -9.353-2 2.06-2-2.844-3
  .
  .
  .
  .
  .

GRID     2025       7.7128-2-3.491-22.3149-3
GRID     2026       7.215-2-3.654-22.4226-3
GRID     2027       6.714-2-3.806-22.5236-3
GRID     2028       6.21-2-3.948-22.6178-3
GRID     2029       .0570313 -4.08-2 2.705-3
GRID     2030       5.1936-2-4.201-22.7851-3
GRID     2031       4.6816-2 -4.31-2 2.858-3
GRID     2032       4.1673-2-4.409-22.9235-3
GRID     2033       3.651-2-4.497-22.9816-3
GRID     2034       .0313282-4.573-23.0321-3
GRID     2035       2.6131-2-4.638-2 3.075-3
GRID     2036       2.0921-2-4.691-23.1102-3
GRID     2037       1.57-2-4.732-23.1377-3
GRID     2038       1.0471-2-4.762-23.1573-3
GRID     2039       5.237-3 -4.78-23.1691-3
$$
$$-------------------------------------------------------------------------------------$
$$          Group Definitions                                                          $
$$-------------------------------------------------------------------------------------$
$$
$$-------------------------------------------------------------------------------------$
$$   HyperMesh name information for generic property collectors                        $
$$-------------------------------------------------------------------------------------$
$$
$$-------------------------------------------------------------------------------------$
$$   Property Definition for 1-D Elements                                              $
$$-------------------------------------------------------------------------------------$
$HMNAME BEAMSECTCOLS          1"beamsectcol1"
$HMNAME BEAMSECTS
$       1    1"beamsection1"
$       2    7   1    0   1.0   1.0   0.0   0.0   0.0
$       0.0  1
$HMNAME BEAMSECTS  BEAMSECTIONSTANDARD      11    1    0Rod
```

```
$HMNAME BEAMSECTS  BEAMSECTIONSTANDARD  PARAMETERS
0.0005  0.0005  10.0
$HMNAME BEAMSECTS  END
$$
$$ CROD Elements
$$
CROD       1      1      1      2
CROD       2      1      2      3
CROD       3      1      1      3
CROD       4      1      4      1
CROD       5      1      4      5
CROD       6      1      3      5
CROD       7      1      6      4
CROD       8      1      6      7
CROD       9      1      5      7
CROD      10      1      8      6
 .
 .
 .
 .
CROD     3751     1    2020   1993
CROD     3752     1    2021   1994
CROD     3753     1    2022   1995
CROD     3754     1    2023   1996
CROD     3755     1    2024   1997
CROD     3756     1    2025   1998
CROD     3757     1    2026   1999
CROD     3758     1    2027   200
CROD     3759     1    2028   2001
CROD     3760     1    2029   2002
CROD     3761     1    2030   2003
CROD     3762     1    2031   2004
CROD     3763     1    2032   2005
CROD     3764     1    2033   2006
CROD     3765     1    2034   2012
CROD     3766     1    2035   2011
CROD     3767     1    2036   2010
CROD     3768     1    2037   2009
CROD     3769     1    2038   2008
CROD     3770     1    2039   2007
$
$HMMOVE     2
$        1THRU      3770
$$
$$-------------------------------------------------------------------------------- ----$
```

```
$$   HyperMesh name and color information for generic components        $
$$-------------------------------------------------------------------- ----$
$HMNAME COMP            1"1"
$HWCOLOR COMP           1    5
$
$HMNAME COMP            2"wire"
$HWCOLOR COMP           2    3
$
$
$HMDPRP
$       1THRU     3770
$
$
$$
$$----------------------------------------------------------------------$
$$   Property Definition for Surface and Volume Elements                $
$$--------------------------------------------------------------------$
$$
$$ PROD Data
$
$HMNAME PROP            1"rod" 3
$HWCOLOR PROP           1    6
$HMBEAMSEC PRODASSOC        1    1
PROD      1    0 7.854 79.817-14
$$-----------------------------------------------------------
$$ HYPERMESH TAGS
$$-----------------------------------------------------------
$$BEGIN TAGS
$$END TAGS
$$
$$----------------------------------------------------------------------$
$$   HyperMesh name information for generic materials          $
$$--------------------------------------------------------------------$
$$
$$--------------------------------------------------------------------$
$$            Material Definition Cards              $
$$--------------------------------------------------------------------$
$$
$$--------------------------------------------------------------------$
$$    Loads and Boundary Conditions              $
$$-------------------------------------------------------------------- $
$$
$$HyperMesh name and color information for generic loadcollectors
$$
$$
ENDDATA
```

C.2 Format of NEC Output File for the Monostatic RCS of NASA Almond (Horizontal Polarization)

```
*********************************************
NUMERICAL ELECTROMAGNETICS CODE (NEC)
*********************************************

- - - - COMMENTS - - - -

MONOSTATIC RCS OF METALLIC NASA ALMOND

FREQUENCY=1.19 GHZ; WAVELENGTH= 0.25 M

ANGLES OF INCIDENCE: THETA=90 DEGREE; PHI=0 TO 180 DEGREE; ETA=90

ANGLES OF SCATTERING: THETA=90 DEGREE; PHI=0 TO 180 DEGREE

- - - STRUCTURE SPECIFICATION - - -
```

COORDINATES MUST BE INPUT IN METERS OR BE SCALED TO METERS BEFORE STRUCTURE INPUT IS ENDED

WIRE NO.	X1	Y1	Z1	X2	Y2	Z2	NO. of RADIUS	FIRST SEG.	LAST SEG.	TAG SEG.	NO.
1	-0.1046	0.0048	-0.0007	-0.1052	0	0	0.0005	1	1	1	1
2	-0.1052	0	0	-0.1046	0.0049	-0.0003	0.0005	2	2	2	1
3	-0.1046	0.0048	-0.0007	-0.1046	0.0049	-0.0003	0.0005	3	3	3	1
4	-0.1029	0.0094	-0.0013	-0.1046	0.0048	-0.0007	0.0005	4	4	4	1
5	-0.1029	0.0094	-0.0013	-0.103	0.0096	-0.0006	0.0005	5	5	5	1
6	-0.1046	0.0049	-0.0003	-0.103	0.0096	-0.0006	0.0005	6	6	6	1
7	-0.1003	0.0135	-0.0019	-0.1029	0.0094	-0.0013	0.0005	7	7	7	1
8	-0.1003	0.0135	-0.0019	-0.1006	0.014	-0.0009	0.0005	8	8	8	1

9	-0.103	0.0096	-0.0006	-0.1006	0.014	-0.0009	0.0005	1	9	9	1
10	-0.0971	0.0173	-0.0024	-0.1003	0.0135	-0.0019	0.0005	1	10	10	1
·											
·											
·											
3760	0.057	-0.0408	0.0027	0.0568	-0.0385	0.0053	0.0005	1	3760	3760	1
3761	0.0519	-0.042	0.0028	0.0517	-0.0396	0.0055	0.0005	1	3761	3761	1
3762	0.0468	-0.0431	0.0029	0.0466	-0.0406	0.0056	0.0005	1	3762	3762	1
3763	0.0417	-0.0441	0.0029	0.0414	-0.0416	0.0057	0.0005	1	3763	3763	1
3764	0.0365	-0.045	0.003	0.0363	-0.0424	0.0059	0.0005	1	3764	3764	1
3765	0.0313	-0.0457	0.003	0.0312	-0.0431	0.006	0.0005	1	3765	3765	1
3766	0.0261	-0.0464	0.0031	0.026	-0.0437	0.006	0.0005	1	3766	3766	1
3767	0.0209	-0.0469	0.0031	0.0208	-0.0442	0.0061	0.0005	1	3767	3767	1
3768	0.0157	-0.0473	0.0031	0.0156	-0.0446	0.0062	0.0005	1	3768	3768	1
3769	0.0105	-0.0476	0.0032	0.0104	-0.0449	0.0062	0.0005	1	3769	3769	1
3770	0.0052	-0.0478	0.0032	0.0052	-0.045	0.0062	0.0005	1	3770	3770	1

TOTAL　SEGMENTS USED = 3770　　NO. SEG. IN A SYMMETRIC　CELL= 3770　　SYMMETRY FLAG = 0

- MULTIPLE WIRE JUNCTIONS

JUNCTION SEGMENTS (FOR END 1, + FOR END 2)

1	-1	-3	4	378	426-	428	590-	872
2	1	-2	-213	589	810	1044	1284	-1349
3	2	3	-6	165-	214-	3104		
4	-4	-5	7	166				
5	5	6	-9	-215				
6	-7	-8	10	168				
7	8	9	-12-	217				

---- SEGMENTATION DATA ----

COORDINATES IN METERS

I+ AND I- INDICATE THE SEGMENTS BEFORE AND AFTER I

SEG. NO.	COORDINATES OF SEG.CENTER			SEG. LENGTH	ORIENTATION ANGLES		WIRE RADIUS	I-	I	I+	TAG NO.
	X	Y	Z		ALPHA	BETA					
8								-10	-11	13	170
9								11	12	-15	-219
10								-13	-14	16	172
· · ·											
1695								3708	-3709	-3710	-3766
1696								3710	-3711	-3712	-3767
1697								3712	-3713	-3714	-3768
1698								3714	-3715	-3716	-3769
1699								3716	-3717	-3718	-3770
1	-0.10490	0.00240	-0.00035	0.00489	8.23396	-97.12502	0.00050	-3	1	2	1
2	-0.10490	0.00245	-0.00015	0.00495	-3.47762	83.01894	0.00050	-213	2	-3	1
3	-0.10460	0.00485	-0.00050	0.00041	75.96376	90.00000	0.00050	4	3	6	1
4	-0.10375	0.00710	-0.00100	0.00494	6.97531	-110.28256	0.00050	-5	4	-378	1
5	-0.10295	0.00950	-0.00095	0.00073	72.28453	116.56505	0.00050	7	5	-6	1
· · ·											
3765	0.03125	-0.04440	0.00450	0.00397	49.06466	92.20260	0.00050	3706	3765	-3661	1
3766	0.02605	-0.04505	0.00455	0.00396	47.02582	92.12110	0.00050	3708	3766	-3659	1
3767	0.02085	-0.04555	0.00460	0.00404	47.99326	92.12110	0.00050	3710	3767	-3657	1
3768	0.01565	-0.04595	0.00465	0.00411	48.92574	92.12110	0.00050	3712	3768	-3655	1

```
3769    0.01045  -0.04625  0.00470    47.99326    92.12110    0.00050    3714  3769 -3653    1
3770    0.00520  -0.04640  0.00470    46.97493    90.00000    0.00050    3716  3770 -3651    1

***** DATA CARD NO. 1  FR  0   1   0   0 0 1.19000E+03 0.00000E+00 0.00000E+00 0.00000E+00 0.00000E+00
***** DATA CARD NO. 2  EX  1   1  181   0 9.00000E+01 0.00000E+00 9.00000E+01 0.00000E+00 0.00000E+00 0.00000E+00
***** DATA CARD NO. 3  RP  0   1  181 1000 9.00000E+01 0.00000E+00 0.00000E+00 1.00000E+00 0.00000E+00 0.00000E+00
```

- - - - - FREQUENCY - - - - - -

FREQUENCY= 1.1900E+03 MHZ
WAVELENGTH= 2.5193E-01 METERS

APPROXIMATE INTEGRATION EMPLOYED FOR SEGMENTS MORE THAN 1.000 WAVELENGTHS APART

- - - STRUCTURE IMPEDANCE LOADING - - -

THIS STRUCTURE IS NOT LOADED

- - - ANTENNA ENVIRONMENT - - -

FREE SPACE

- - - MATRIX TIMING - - -

FILL= 23.692 SEC., FACTOR= 30.088 SEC.

- - EXCITATION - - -

PLANE WAVE   THETA= 90.00 DEG, PHI= 0.00 DEG, ETA= 90.00 DEG, TYPE -LINEAR= AXIAL RATIO= 0.000

- - - CURRENTS AND LOCATION - - -

DISTANCES IN WAVELENGTHS

SEG. NO.	TAG NO.	COORD. OF SEG. CENTER X	Y	Z	SEG. LENGTH	- - - CURRENT (AMPS) - - - REAL	IMAG.	MAG.	PHASE
1	1	-0.4164	0.0095	-0.0014	0.01940	7.3112E-05	-1.4301E-04	1.6062E-04	-62.922
2	1	-0.4164	0.0097	-0.0006	0.01963	1.2360E-04	-2.4493E-04	2.7436E-04	-63.222
3	1	-0.4152	0.0193	-0.0020	0.00164	2.1327E-02	-3.5383E-02	4.1313E-02	-58.921
4	1	-0.4118	0.0282	-0.0040	0.01961	-1.0609E-04	9.5121E-05	1.4249E-04	138.121
5	1	-0.4086	0.0377	-0.0038	0.00292	4.0604E-05	-3.6441E-05	5.4558E-05	-41.907
.									
.									
3765	1	0.1240	-0.1762	0.0179	0.01576	1.4959E-06	4.8635E-07	1.5730E-06	18.010
3766	1	0.1034	-0.1788	0.0181	0.01573	9.6220E-07	1.1785E-06	1.5214E-06	50.770
3767	1	0.0828	-0.1808	0.0183	0.01603	1.1274E-06	1.4822E-07	1.1371E-06	7.490
3768	1	0.0621	-0.1824	0.0185	0.01632	1.4785E-06	-5.8145E-07	1.5887E-06	-21.468
3769	1	0.0415	-0.1836	0.0187	0.01603	1.1100E-06	4.1951E-07	1.1866E-06	20.703
3770	1	0.0206	-0.1842	0.0187	0.01629	1.1727E-06	4.9439E-07	1.2726E-06	22.859

- - - RADIATION PATTERNS - - -

- - ANGLES - - THETA DEGREES	PHI DEGREES	- POWER GAINS - VERT. DB	HOR. DB	TOTAL DB	- - POLARIZATION - - - AXIAL RATIO	TILT DEG.	SENSE	- - - E(THETA) - - - MAGNITUDE VOLTS/M	PHASE DEGREES	- - - E(PHI) - - - MAGNITUDE VOLTS/M	PHASE DEGREES
90.00	0.00	-98.83	-15.01	-15.01	0.00005	-90.00	LEFT	8.13304E-07	77.99	1.26219E-02	-159.76
90.00	1.00	-98.54	-15.01	-15.01	0.00006	-90.00	LEFT	8.40367E-07	83.61	1.26202E-02	-159.75
90.00	2.00	-98.22	-15.02	-15.02	0.00006	-90.00	LEFT	8.72519E-07	88.90	1.26150E-02	-159.72
90.00	3.00	-97.86	-15.02	-15.02	0.00007	-90.00	LEFT	9.09425E-07	93.83	1.26062E-02	-159.67
90.00	4.00	-97.47	-15.03	-15.03	0.00007	-90.00	LEFT	9.50728E-07	98.40	1.25939E-02	-159.60
90.00	5.00	-97.07	-15.04	-15.04	0.00008	-90.00	LEFT	9.96065E-07	102.61	1.25779E-02	-159.51
.											
.											
90.00	175.00	-97.32	-13.44	-13.44	0.00003	-90.00	RIGHT	9.67618E-07	-74.39	1.51200E-02	131.43
90.00	176.00	-96.84	-13.40	-13.40	0.00003	-90.00	RIGHT	1.02308E-06	-75.99	1.52010E-02	131.76
90.00	177.00	-96.38	-13.36	-13.36	0.00003	-90.00	RIGHT	1.07852E-06	-77.63	1.52644E-02	132.02
90.00	178.00	-95.94	-13.33	-13.36	0.00004	-90.00	RIGHT	1.13417E-06	-79.30	1.53098E-02	132.21
90.00	179.00	-95.52	-13.32	-13.32	0.00004	-90.00	RIGHT	1.19023E-06	-80.99	1.53371E-02	132.32
90.00	180.00	-95.12	-13.31	-13.31	0.00005	-90.00	RIGHT	1.24687E-06	-82.67	1.53462E-02	132.35

- - - EXCITATION - - -

PLANE WAVE   THETA= 90.00 DEG, PHI= 1.00 DEG, ETA= 90.00 DEG, TYPE -LINEAR= AXIAL RATIO= 0.000

- - - CURRENTS AND LOCATION - - -

DISTANCES IN WAVELENGTHS

SEG. NO.	TAG NO.	COORD. OF SEG. CENTER X	Y	Z	SEG. LENGTH	REAL	IMAG.	MAG.	PHASE
						- - - CURRENT (AMPS) - - -			
1	1	-0.4164	0.0095	-0.0014	0.01940	9.0643E-05	-1.9871E-04	2.1841E-04	-65.479
2	1	-0.4164	0.0097	-0.0006	0.01963	1.5135E-04	-3.3207E-04	3.6494E-04	-65.497
3	1	-0.4152	0.0193	-0.0020	0.00164	2.4324E-02	-4.2742E-02	4.9179E-02	-60.356
4	1	-0.4118	0.0282	-0.0040	0.01961	-1.1868E-04	1.3407E-04	1.7906E-04	131.516
5	1	-0.4086	0.0377	-0.0038	0.00292	4.5117E-05	-5.0273E-05	6.7549E-05	-48.094
. . .									
3765	1	0.1240	-0.1762	0.0179	0.01576	1.4994E-06	4.7225E-07	1.5720E-06	17.482
3766	1	0.1034	-0.1788	0.0181	0.01573	9.8780E-07	1.1565E-06	1.5209E-06	49.498
3767	1	0.0828	-0.1808	0.0183	0.01603	1.1270E-06	1.4300E-07	1.1360E-06	7.231
3768	1	0.0621	-0.1824	0.0185	0.01632	1.4557E-06	-5.7623E-07	1.5656E-06	-21.596
3769	1	0.0415	-0.1836	0.0187	0.01603	1.1175E-06	4.0860E-07	1.1898E-06	20.084
3770	1	0.0206	-0.1842	0.0187	0.01629	1.1843E-06	4.8065E-07	1.2781E-06	22.090

- - - RADIATION PATTERNS - - -

- - ANGLES - - THETA DEGREES	PHI DEGREES	- POWER GAINS - VERT. DB	HOR. DB	TOTAL DB	- - POLARIZATION - - AXIAL RATIO	TILT DEG.	SENSE	- - - E(THETA) - - - MAGNITUDE VOLTS/M	PHASE DEGREES	- - - E(PHI) - - - MAGNITUDE VOLTS/M	PHASE DEGREES
90.00	0.00	-94.67	-15.01	-15.01	0.00010	-90.00	LEFT	1.31243E-06	93.28	1.26201E-02	-159.74
90.00	1.00	-94.39	-15.01	-15.01	0.00010	-90.00	LEFT	1.35615E-06	96.41	1.26286E-02	-159.88
90.00	2.00	-94.10	-15.00	-15.00	0.00011	-90.00	LEFT	1.40115E-06	99.40	1.26335E-02	-159.99
90.00	3.00	-93.82	-15.00	-15.00	0.00011	-90.00	LEFT	1.44748E-06	102.24	1.26350E-02	-160.09
90.00	4.00	-93.54	-15.00	-15.00	0.00012	-90.00	LEFT	1.49520E-06	104.94	1.26328E-02	-160.16
90.00	5.00	-93.26	-15.01	-15.01	0.00012	-90.00	LEFT	1.54433E-06	107.50	1.26270E-02	-160.22

90.00	175.00	-93.26	-13.60	0.00006	-90.00	RIGHT	1.54342E-06	-81.99	1.48479E-02	131.57
90.00	176.00	-92.95	-13.52	0.00006	-89.99	RIGHT	1.60101E-06	-82.77	1.49813E-02	131.88
90.00	177.00	-92.64	-13.46	0.00006	-89.99	RIGHT	1.65837E-06	-83.65	1.50970E-02	132.10
90.00	178.00	-92.35	-13.40	0.00007	-89.99	RIGHT	1.71563E-06	-84.60	1.51948E-02	132.25
90.00	179.00	-92.06	-13.35	0.00007	-89.99	RIGHT	1.77295E-06	-85.60	1.52744E-02	132.32
90.00	180.00	-91.78	-13.32	0.00008	-89.99	RIGHT	1.83045E-06	-86.65	1.53358E-02	132.32

- - - EXCITATION - - -

PLANE WAVE   THETA= 90.00 DEG, PHI= 180.00 DEG, ETA= 90.00 DEG, TYPE -LINEAR= AXIAL RATIO= 0.000

- - - CURRENTS AND LOCATION - - -

DISTANCES IN WAVELENGTHS

| SEG. NO. | TAG NO. | COORD. OF SEG. CENTER | | | SEG. LENGTH | - - - CURRENT (AMPS) - - - | | | |
		X	Y	Z		REAL	IMAG.	MAG.	PHASE
1	1	-0.4164	0.0095	-0.0014	0.01940	4.9431E-05	1.0137E-04	1.1278E-04	64.005
2	1	-0.4164	0.0097	-0.0006	0.01963	8.3261E-05	1.7507E-04	1.9386E-04	64.564
3	1	-0.4152	0.0193	-0.0020	0.00164	1.4653E-02	2.5034E-02	2.9007E-02	59.659
4	1	-0.4118	0.0282	-0.0040	0.01961	-7.9548E-05	-4.0650E-05	8.9333E-05	-152.933
.									
.									
3767	1	0.0828	-0.1808	0.0183	0.01603	-7.3895E-07	5.3039E-08	7.4085E-07	175.895
3768	1	0.0621	-0.1824	0.0185	0.01632	-1.2055E-07	6.4141E-07	6.5264E-07	100.644
3769	1	0.0415	-0.1836	0.0187	0.01603	-1.0355E-06	-3.2838E-07	1.0863E-06	-162.405
3770	1	0.0206	-0.1842	0.0187	0.01629	-1.0149E-06	-5.3487E-07	1.1472E-06	-152.210

- - - RADIATION PATTERNS - - -

- - ANGLES - -		- POWER GAINS -			- - POLARIZATION - - -			- - - E(THETA) - - -		- - - E(PHI) - - -	
THETA DEGREES	PHI DEGREES	VERT. DB	HOR. DB	TOTAL DB	AXIAL RATIO	TILT DEG.	SENSE	MAGNITUDE VOLTS/M	PHASE DEGREES	MAGNITUDE VOLTS/M	PHASE DEGREES
90.00	0.00	-101.69	-13.37	-13.37	0.00004	90.00	RIGHT	5.85155E-07	-144.18	1.52416E-02	132.84
90.00	1.00	-101.25	-13.38	-13.38	0.00004	-90.00	RIGHT	6.15410E-07	-135.38	1.52320E-02	132.81
90.00	2.00	-100.67	-13.39	-13.39	0.00004	-90.00	RIGHT	6.57608E-07	-127.40	1.52031E-02	132.71
90.00	3.00	-100.01	-13.42	-13.42	0.00004	-90.00	RIGHT	7.10271E-07	-120.36	1.51552E-02	132.55
90.00	177.00	-98.02	-16.23	-16.23	0.00003	90.00	RIGHT	8.92134E-07	163.46	1.09743E-02	142.49
90.00	178.00	-98.53	-16.22	-16.22	0.00002	90.00	RIGHT	8.41567E-07	158.32	1.09817E-02	142.51
90.00	179.00	-99.02	-16.22	-16.22	0.00001	90.00	RIGHT	7.95994E-07	152.72	1.09862E-02	142.52
90.00	180.00	-99.46	-16.22	-16.22	0.00000	90.00	LINEAR	7.56404E-07	146.63	1.09878E-02	142.53

***** DATA CARD NO. 4 EN 0 0 0 0 0 0.00000E+00 0.00000E+00 0.00000E+00 0.00000E+00 0.00000E+00

RUN TIME = 152.380

Author Index

© The Author(s), under exclusive license to Springer Nature Singapore Pte Ltd. 2021
V. Joy et al., *Fundamentals of RCS Prediction Methodology using Parallelized Numerical
Electromagnetics Code (NEC) and Finite Element Pre-processor*,
SpringerBriefs in Computational Electromagnetics,
https://doi.org/10.1007/978-981-15-7164-0

Subject Index

© The Author(s), under exclusive license to Springer Nature Singapore Pte Ltd. 2021   83
V. Joy et al., *Fundamentals of RCS Prediction Methodology using Parallelized Numerical
Electromagnetics Code (NEC) and Finite Element Pre-processor*,
SpringerBriefs in Computational Electromagnetics,
https://doi.org/10.1007/978-981-15-7164-0

Printed in the United States
By Bookmasters